The Higher Learning and High Technology

SUNY Series
FRONTIERS IN EDUCATION
Philip G. Altbach, Editor

The Frontiers in Education Series features and draws upon a range of disciplines and approaches in the analysis of educational issues and concerns, helping to reinterpret established fields of scholarship in education by encouraging the latest synthesis and research.

Other books in this series include:

Class, Race, and Gender in American Education
—Lois Weis (ed.)

Excellence and Equality: A Qualitatively Different Perspective on Gifted and Talented Education
—David M. Fetterman

Change and Effectiveness in Schools: A Cultural Perspective
—Gretchen B. Rossman, H. Dickson Corbett, and William A. Firestone

The Curriculum: Problems, Politics, and Possibilities
—Landon E. Beyer and Michael W. Apple (eds.)

The Character of American Higher Education and Intercollegiate Sport
—Donald Chu

Crisis in Teaching: Perspectives on Current Reforms
—Lois Weis, Philip G. Altbach, Gail P. Kelly, Hugh G. Petrie, and Sheila Slaughter (eds.)

The High-Status Track: Studies of Elite Schools and Stratification
—Paul William Kingston and Lionel S. Lewis (eds.)

The Economics of American Universities: Management, Operations, and Fiscal Environment
—Stephen A. Hoenack and Eileen L. Collins (eds.)

Dropouts from Schools: Issues, Dilemmas and Solutions
—Lois Weis, Eleanor Farrar and Hugh Petrie

Religious Fundamentalism and American Education: The Battle for the Public Schools
—Eugene F. Provenzo, Jr.

The Higher Learning and High Technology

Dynamics of Higher Education Policy Formation

SHEILA SLAUGHTER

State University of New York Press

Published by
State University of New York Press, Albany

© 1990 State University of New York

All rights reserved

Printed in the United States of America

No part of this book may be used or reproduced
in any manner whatsoever without written permission
except in the case of brief quotations embodied in
critical articles and reviews.

For information, address State University of New York
Press, State University Plaza, Albany, N.Y., 12246

Library of Congress Cataloging in Publication Data

Slaughter, Sheila.
   The higher learning and high technology: dynamics of higher
education policy formation/Sheila Slaughter.
     p.   cm.—(SUNY series, frontiers in education)
  Includes index.
  ISBN 0-7914-0048-4.   ISBN 0-7914-0049-2 (pbk.)
  1. High technology and education—United States. 2. Higher
education and state—United States. 3. Industry and education—
United States. I. Title. II. Series.
LC1087.2.S58   1990
378'.103—dc19

88-30581
CIP

10 9 8 7 6 5 4 3 2 1

In memory of my mother, Evelyn Kastner Slaughter, and
in hope for the future of my daughter, Jessica Slaughter McVey

"As bearing on the case of the American universities, it should be called to mind that the businessmen of this country, as a class, are of a notably conservative habit of mind."

> Thorstein Veblen, *The Higher Learning in America: A Memorandum on the Conduct of Universities by Business Men,* 1918.

Contents

List of Tables	ix
Acknowledgements	xi
Chapter 1 The Higher Education Policy Literature	1
Chapter 2 Policy Issues Past and Present	27
Chapter 3 Social Location of Corporate and University Leaders	61
Chapter 4 University Presidents and Public Policy	97
Chapter 5 Bases of Corporate Interest in Higher Education	143
Chapter 6 The Business-Higher Education Forum Reports: Corporate and Campus Cooperation	171
Chapter 7 The Business-Higher Education Forum Reports: New Issues and Traditional Alliances	189
Chapter 8 Class, State, Ideology, and Higher Education	217
Notes	245
Index	283

List of Tables

3.1.	Leaders' B.A. by Carnegie Classification and Control	88
3.2.	Leaders' M.A. by Carnegie Classification and Control	90
3.3.	Leaders' Ph.D. by Carnegie Classification and Control	91
3.4.	Leaders' Highest Degree by Field of Specialization	92
3.5.	Leaders' Military Service	92
3.6.	Leaders' Career Patterns	93
3.7.	Number of Institutions in Leaders' Career Path	93
3.8.	Leaders' Representation on Corporate Boards	94
3.9.	Leaders' Representation on Policy Institutes or Forums	94
3.10.	Leaders' Trade and Professional Associations	95
3.11.	Leaders' Representation in Social Clubs	95
4.1.	Associations Officially Represented by AAU Presidents in Their Congressional Testimony	100
4.2.	Instances of Presidential Testimony	102
4.3.	Congressional Committees Before Which Presidents Testified	110
4.4.	Topics on Which AAU Presidents Spoke in Congressional Testimony	112

4.5. Presidential Themes	118
5.1. Corporations Participating in the Business-Higher Education Forum for Two or More Years and Their Representatives	147
5.2. Wholly Owned Consolidated Subsidiaries of Forum Corporations	150
5.3. Percentage of Net Sales Devoted to Research and Development by Forum Corporations	153
5.4. Corporate Gifts to Higher Education	156

Acknowledgements

Many people helped me through this manuscript in a variety of ways. Michael Olivas appointed me as a research fellow at the Institute for Higher Education Law and Governance at the University of Houston, which gave me a precious year off, with time to read and think. Lois Patton, editor-in-chief at SUNY Press, offered strong editorial commitment to the project. As always, my colleagues and friends made invaluable contributions, reading the manuscript in various drafts and offering insightful criticism. David Noble, at Drexel, was a faithful reader. Philip Altbach and Lois Weis, at SUNY-Buffalo, have read my work over many years, and as always, contributed richly to our ongoing dialogue about the relationship between knowledge and power. Michael Useem, Boston University, read and commented on Chapter 3, as did Neil Fligstein, University of Arizona, on Chapter 5. My new colleagues in Higher Education at the University of Arizona, Clifton Conrad, now at Wisconsin, Larry Leslie, and Gary Rhoades, came to know more about my work than they might have anticipated, reading, commenting on, and often disagreeing with what I wrote as it came out, chapter by chapter. The exchanges we had over this manuscript created the engagement that leads to productive intellectual camaraderie. Although the critiques offered by my colleagues helped me enormously, they are not in any way implicated in the arguments advanced here.

CHAPTER 1

The Higher Education Policy Literature

The central problem this book addresses is the national policy formation process in higher education. Recent scholarship has focused primarily on policy formation problems that relate to rapidly growing postsecondary systems, on the one hand, and problems stemming from fiscal crisis, on the other.[1] What has been overlooked are new configurations of actors and interests that are emerging as a mature postsecondary education system interacts with changing structural conditions in the political economy. Although the central actors have not changed—state, industry, higher education associations and philanthropic foundations, faculty, and those portions of the public seeking a higher education and the benefits that stem from it—their complexity and the relationships between them have. Together, these subtle shifts have created new patterns of policy formation, and, of course, new policy agendas.

Historically, the state, particularly the federal government, has been seen as either the font of higher education funding, whether for science or system expansion, or as an intrusive force, interfering with the development of professional and scientific expertise and constraining the free market in faculty.[2] Although these are indeed characteristics of government action, scholars have begun to recognize that the state is intertwined with and perhaps inseparable from postsecondary education. It becomes ever more apparent that the state is not monolithic, that various branches, levels, and component parts of the state respond to different, sometimes competing or conflicting constituencies, and, in some instances, that the state is able to order constructively the claims, interests, and energies of players in the higher-education policy formation process.[3] The state, then, is no longer simply the source of monies or a policing agency, but is simultaneously a multifaceted resource, the arena in which policy formation is played out, and an actor in its own right, with an often unpredictable agenda.

After World War II, the influence of private industry on the higher education policy formation process waned. The significance of private giving was diminished by government spending. However, as worldwide economic competition increases and as the place of the United States in the global economy becomes more and more tenuous, American corporations have come together to give cohesive voice to their needs. They have argued convincingly that all policy has to subordinate itself to America's regaining its competitive edge in the global market. The vehicle for economic success is high technology.[4] For universities, this means emphasizing "advanced applied" technology, technology transfer, and the training of a competitive scientific and professional labor force.[5] Because of greater organization and effort on the part of the corporate community and the change in the U.S. position in the global market, the voice of the corporate community is coming to be more closely heeded in the higher education policy-making process.

In the immediate postwar period, philanthropic foundations were a strong voice in higher education policy formation, whereas higher education associations were inadequately organized and relatively weak.[6] When confronted with the crisis in funding at the federal level after the Vietnam War and, somewhat later, by fiscal crises in various of the several states, higher education associations organized more effectively. Given the increased number of postsecondary schools and the ever-greater amount of resources they came to command, the stake of these institutions in the outcomes of the policy process grew, and their cumulative efforts have come to exceed those of foundations.[7] Although foundations continue to address the systemic problems faced by postsecondary education, they are one voice among many rather than *primus inter pares*. The leaders of research universities, through their national associations, have become especially aggressive with regard to securing funds for their institutions. More specifically, they have begun to emphasize university-industry partnerships developed around the exploitation of high technology in which the university holds intellectual property rights and sometimes equity positions.

Traditionally, select faculty in the physical and natural sciences—organized in quasi-public bodies such as the American Association for the Advancement of Science, the National Academy of Science, the National Research Council, and the associations of learned disciplines—worked with the White House and the federal mission agencies, particularly those concerned with national defense and health, to shape science policy.[8] Although faculty still maintain these relationships, other avenues for influencing national science policy are opening up. Science faculty, often in interdisciplinary groups, are forming alliances with industrial corporations, patent officers, and venture capitalists to establish viable high technology corporations. Frequently their efforts are sponsored by their universities.

In the 1960s and 1970s, those diverse segments of the public concerned with higher education were able to join together around issues such as increased access, system expansion, and keeping tuition costs low. As many of these demands were met at some level of the system, the cohesion of these groups diminished. Currently, popular calls for reform of higher education policy are fragmented, and there seems to be no clear policy agenda.

I argue, then, that over time the relationships of the central actors in the national higher education policy formation process have shifted. Rather than the federal government, together with the scientific community and foundations, having the dominant voice, the corporate community, in partnership with leaders of research universities, has become more vocal and taken a more active position. I do not want to suggest that traditional relationships important to the policy formation process have irreparably broken down or have been completely displaced, but I do think there are indications that these relationships are being refashioned subtly and, in some instances, superseded by organizations and alliances better able to sustain the flow of resources to higher education in an era when economic competitiveness is central to all policy makers' agendas.

To explore these changes in the policy formation process, I looked for a vehicle that would allow me to examine the shifting balance of policy formation power. A number of choices presented themselves, since many organizations concerned with economic competitiveness and higher education have emerged in the past decade, nationally, regionally, and locally: the Business–Higher Education Forum, the Carnegie Forum on Education and Economy, the President's Commission on Industrial Competitiveness, the Government-University-Industry Research Roundtable, the Massachusetts High Technology Commission, and the New Jersey Governor's Commission on Science and Technology are a few among many. I finally settled on the Business–Higher Education Forum.

I chose the Forum for several reasons. The Business–Higher Education Forum, started in 1978, preceded most of the new industry-university organizations. The Forum is national in scope, and was initially confined to representatives of the two groups I see as central to present policy formation dynamics: the chief executive officers (CEOs) of U.S.-based multinational corporations, and the presidents of prestigious research universities. The Forum was conceived by the American Council of Education (ACE), always a force to be reckoned with in the higher education policy formation process. The majority of the presidents in the Forum represent institutions that are members of the Association of American Universities (AAU). The AAU, once primarily a president's club, has become more and more active in recent years, and typifies the increasingly aggressive stance that presidents of elite research universities are taking toward the policy process. The CEOs too are active in other policy organizations that are influential in shaping policy. Many CEOs are members of the Committee on Economic Development and the Business Roundtable, which are policy organizations that figure in most business community deliberations and that also address problems pertaining to higher education. The Forum, then, seemed to be a confluence of groups that are increasingly active in the policy process and an appropriate subject for the case study which is at the heart of this volume. Although the Forum seemed particularly suited to shed light on the questions I am asking about national policy formation, I think that many other organizations that are concerned with industry-university-government relations, whether old or new, would tell a similar story.

The specific content and various methodological approaches used in this book are discussd later in this chapter. I would, however, like to say a word about my theoretical framework at this point. I see the higher education policy literature as undertheorized. Indeed, most of this literature has been written from an atheoretical perspective which assumes, but does not articulate, a pluralist or neopluralist framework. Because I think theory and theory construction are important to understanding the policy process, particularly when we are trying to understand change, glimpse the future, and influence policy, I will try to articulate the theoretical questions addressed by the various data sets presented in this book, and view them through competing lenses. The lenses I have chosen are pluralism, neo-Marxism, and corporatism. Pluralism is important because it informs the body of the higher education policy literature. Neo-Marxism is salient because it raises the critical questions pluralism has left unasked, especially with regard to class, state, ideology, and the disposition of power in the policy process.

Corporatism is significant because it addresses changes in the political process unanticipated by neo-Marxists.

In this first chapter, I review the higher-education policy literature, which I take to include work about the policy process and policy issues, the literature on the relation of business to the postsecondary sector, and the science policy literature. I see these literatures as relevant because they shed light on the policy formation process, which transcends party politics and the legislative arena. Next, I describe the Business–Higher Education Forum and argue for its centrality to the policy process that is emerging in the late twentieth century. Finally, I speak to the theoretical issues, content, and methods of each of the chapters constituting this volume.

Policy Literature in Postsecondary Education

As students of policy literature note,

> the defining characteristics of a discipline 'include a general body of knowledge ... a specialized vocabulary and a generally accepted basic literature ... some generally accepted body of theory and some generally understood techniques for theory testing and revision ... a generally agreed upon methodology ... and recognized techniques for replication and revalidation of research and scholarship.' According to several observers, the study of higher education politics and policy is lacking in all of these respects.[9]

From a pluralist perspective, this literature is not well defined in the terms of traditional scholarly disciplines. However, when viewed through a neo-Marxian lens, it has a certain substantive coherence. If the central question is cast as, who benefits in terms of groups' social location (class) vis-á-vis the relations of production (economy) then the majority of the policy literature is focused on issues of access and equity. When set in historical context, considerations of access and equity are frequently followed by literatures raising questions about cost containment, quality, and autonomy, which, in functional terms, frequently serve to protect privilege. The policy literature seems to be informed by a dynamic or dialectic in which unentitled groups press for the expansion of higher education as a way of improving their class position, while representatives of elite institutions, whether private or public, attempt to constrict easy access to maintain control over institutions central to perpetuating class privilege.

Generally, policy has preceded a well-developed literature grounded in research. During World War II, the GI Bill was formulated without extensive reliance on research. It was a response to the unrest anticipated after a long and grueling war animated by powerful social ideals, the overriding one of which was to make the world safe for democracy. Political leaders thought that returning veterans would demand concrete, federally sponsored compensation for their years of service. Rather than redistributing wealth or economic opportunity, the government brought forth educational opportunity, as embodied in the GI Bill. These government initiatives for expanding access were initially prompted by popular demands rather than rational policy analysis performed by experts.[10]

Formal policy literature did not emerge until after the war, and was for some years composed largely of position papers and documents developed by ad hoc commissions appointed by U.S. presidents, by committees and commissions established by the major foundations, and by the higher education associations. This literature reflects enduring aspects of policy dynamics in the United States. Governmental policy initiatives, often shaped in response to popular demands and political events, were developed to solve social, economic, and military problems, or a combination thereof, through education rather than through direct intervention via mechanisms such as regulation, national planning, or redistribution of wealth, goods, and services. Higher education and learned associations, frequently funded by foundations and sometimes aided by business organizations, responded with their own policy positions which asserted their interests, which often ran counter to governmental initiatives.

This dynamic is well illustrated by the various commissions established in the immediate postwar period. In 1947, the President's Commission on Higher Education, popularly known as the Zook Commission, developed a six-volume report which called for a dramatic expansion of higher education, concentrating on the development of statewide systems of tuition-free community colleges as well as federal capital outlay to expand the public sector. The Zook Commission saw the expansion of higher education as a means of calming the fears created by the recession that occurred after World War II and as a means of responding to national needs for a more technically competent and scientifically skilled labor force. The private sector developed an alternative to the Zook Commission proposals. With funding from the Rockefeller Foundation, the AAU created the Committee on Financing Higher Education. Diverging sharply from the Zook Commission, the Committee spoke for providing education only to the most talented, not more than the top 25 percent, against heavily subsidized tuition at state institutions, against direct federal aid to colleges and universities, and against scholarship aid to individual students.[11] In contrast to the Zook Commission, the higher education establishment moved for highly selective expansion that would protect its position.

With variations, this dynamic is played and replayed, dominating the higher education policy literature. The debates over public subsidy for postsecondary education, student aid, and affirmative action provide examples.

In the subsidy debates, ad hoc presidential commissions from the Truman era through the Carter administration advocated increases, usually for public sector institutions and frequently in the form of student aid, as a means of creating access and equity.[12] In many instances, these initiatives were formulated in response to popular demands for increased access and educational opportunity voiced by groups from civil rights activists, women, and minorities to the many middle-class parents who sought upward mobility for their children. Alternative positions were developed by business organizations, foundations, think tanks, and education associations representing elite institutions. Initially, these institutions and organizations argued for curtailing subsidies to public higher education as a way of containing costs and preserving private higher education. The "independent sector," as it came to be called, like the private economic sector, was presented as an unequivocal good, preserving quality, excellence, and educational initiative. For example, the Committee on Economic Development, an organization composed of two hundred members, most of whom are CEOs of large corporations or presidents of universities, argued that states should raise tuitions to cover approximately half of the actual cost of higher education, undercut subsidies for public higher education, and thereby preserve the quality of private higher education by making it cost competitive.[13]

As the public sector rapidly expanded, despite the organizational efforts of the higher education establishment, the private sector changed tactics, moving to include itself in federal subsidies. Foundations and think tanks, along with the higher education associations, sponsored policy documents aimed at shifting subsidies as well as cutting costs. The Carnegie group, through its Commission and Council, began to argue for cost curtailment and preservation of the private sector in the early 1970s.[14] The Brookings Institution, with financial support from the Andrew W. Mellon, Exxon Education, Cleveland, and U.S. Steel foundations, sponsored work such as *Public Policy and Private Higher Education,* which saw the "fundamental policy choice" as "making private colleges financially more like public ones, or public colleges more like private ones."[15] Whether advocating more costly public institutions, less monies for public institutions, or direct subsidy for private institutions, the outcome was generally preservation of the private sector by shifting costs to parents, students, and the taxpaying public.

These issues were not always debated in terms of public versus private. Policy debates were also couched in terms of system rationality versus unfettered access. In this formulation, quality inadvertently became a function of mission and institutional characteristics. In the late 1960s and 1970s, the definition of institutional mission was presented as a means of system rationalization. A broad literature developed around mission definition and system planning. Institutions were classified according to certain characteristics—size, federal research dollars, student test scores, and the like. Probably the most well-known and influential effort in this area was the Carnegie Commission's *Classification of Institutions of Higher Education.*[16] Although policy makers and planners were careful not to equate mission with quality, in point of fact, those institutions with certain designations—flagship, Research I—were most likely to continue receiving the public resources that made quality possible. System rationalization in effect confirmed existing hierarchies of institutional prestige, limiting competition by curbing the aspirations of all institutions. However, in this formulation, established and prestigious public institutions were included along with private. Elite status rather than control designation became the key to differential resource allocation. The long-term result seems to be a stratification system that confirms existing hierarchies of privilege, whether private or public.

The student aid literature reveals a similar dynamic. In response to popular demand and federal initiatives, aid packages were developed for the previously excluded. These were often countered by the higher education establishment, private and public, which offered proposals refashioned to more nearly meet the needs of the existing system. Veterans aid, formulated for a protected class, was offset by National Defense Education Loans, targeted for the talented. All qualified students were included in basic educational opportunity grants of the 1972 Educational Amendments. This legislation was vociferously opposed by the higher education associations, which argued forcefully that federal aid should flow directly to institutions rather than to students.[17] Currently, government efforts are being made to target scarce resources on very needy student populations, and the higher educational establishment is attempting to preserve the student aid and loan programs that have greatly benefited the middle class as college costs have soared.

The affirmative action policy literature provides yet another variant of the general dynamic described. Again, policy preceded the development of a well-articulated literature. Nixon's 1972 Executive Order was not the product of carefully researched policy alternatives; it was a response to political pressure from a variety of groups—civil rights, women's, and minority groups. Policy positions opposing the legislation appeared very quickly after passage. In forums and publications, business groups and blue ribbon higher education commissions frequently came together to argue against the legislation, promising that they could and would do a better job of policing themselves than external agencies.[18] These groups centered their arguments on autonomy. In the case of affirmative action, autonomy meant freedom from state and federal restraints with regard to hiring and firing, promotion and retention. More generally, arguments for autonomy were antiregulation arguments which opposed constraint of managerial privilege whether in the business or higher education sector. An opposing literature, largely, although not exclusively, developed by women and minorities, focused on how to monitor, implement, and assess the impact of regulatory legislation on higher education, usually with an eye to increasing the representation of previously excluded groups.[19]

Of course, not all literature dealing with policy issues is the product of dynamic or dialectic interaction between privileged and unentitled classes. As the postsecondary sector expanded, so did the literature which addressed it. For example, works with policy implications, such as Clark Kerr's *The Uses of the University* and Christopher Jencks and David Riesman's *Academic Revolution,* dealt independently with major structural changes and expansion.[20]

As postsecondary education grew, higher education became a field of study in its own right, with departments in universities as well as professional and scholarly associations, and a series of journals, which further stimulated the authorship and audience for works dealing with colleges and universities.[21] As policy questions became staples for higher education scholars and students, a policy literature began to emerge somewhat independently from government agencies, foundations, business policy formation organizations, and the higher-education associations. Examples are Astin's *Predicting Academic Performance in Colleges,* Gladieux and Wolanin's *Congress and the Colleges* and King's *The Washington Lobbyists.*[22]

The "independent" literature, of course, is permeated with concepts developed in the policy literature discussed earlier. At the present time, it is difficult to distinguish sharply between the several strands. The work of professors speaking as representatives of the postsecondary sector and traditional educational values is often sponsored by the higher education associations and later used in policy documents sponsored by foundations, ad hoc commissions, and the associations. The group that participated in the National Institute of Education (NIE) report, *Involvement in Learning,* provides an example of the complex interaction between government, foundations, associations, and professors of higher education with regard to the production of policy literature. Before becoming a member of the study group that wrote the report, Alexander Astin had written books sponsored jointly, severally, or singly by the Ford Foundation, the Carnegie Commission on Higher Education, the Carnegie Corporation, the National Merit Scholarship Corporation, the National Institute of Health, and the Exxon Foundation. Other members of the NIE group reveal a similar pattern of sponsorship. Howard Bowan's work was supported by the National Council of Churches, Exxon, Ford, the Carnegie Council on Policy Studies in Higher Education, TIAA/CREF, the American Association of Higher Education, the National Institute of Independent Colleges, and the Lilly Foundation. Zelda Gamson's work acknowledges the National Institute for Education, Fund for the Improvement of Postsecondary Education and Exxon; Harold Hodgkinson's mentions Carnegie, Danforth, the Church Society for College Work, Health, Education and Welfare, and the Office of Education. Some of Barbara Lee's work was sponsored by the National Institute of Education and Carnegie, as was Kenneth Mortimer's.[23]

The policy literature developed by scholars such as these fits loosely into what Gladieux and Wolanin have identified as the "higher education policy arena."[24] In many respects this work represents the interests of professors and associations committed to sustaining a fairly autonomous and vigorous postsecondary sector dedicated to traditional academic values. Although a few business organizations are among the sponsors of these professors' work, the "independent" literature may have come to reflect the interests of the higher education establishment, especially those concerned with undergraduate education, and may not speak to the emerging needs of a business community looking to higher education for help in developing high technology as a means of maintaining its place in a highly competitive international economy. If this is the case, the analysis of the policy literature presented here suggests that a new line of policy literature focusing more closely on private sector needs will emerge, a point which will be more fully discussed in the next section.

The critical policy literature is slight. Barbara Ann Scott's *Crisis Management in American Higher Education* is the only work which comes immediately to mind.[25] Her contribution is valuable, especially with regard to emerging institutional and personnel stratification systems. However, Scott conflates elite and neo-Marxian theory, and considers neither science policy nor the state, leaving her critique partial and underdeveloped.

In sum, the higher education policy literature is not as diffuse as it initially appears. Generally, it is ordered by questions of access and equity, which are countered with issues of excellence, quality, and autonomy. Although the literature turns upon who gets what, and according to what principles, certain systemic ordering concepts—such as social class and shared institutional concerns—are by and large ignored. Even the professions and associations that speak for postsecondary organizations and institutions are not regularly considered as having self-serving sectoral interests.

The higher education policy literature very often treats the state, but not usually as something which is problematic. The state is most often seen as an external agency whose only power should be the purse and whose other interventions are illegitimate, as in the literature on regulation. More recently, it has been treated as an administrator or implementor of policy, as in the student aid and affirmative action literatures. Rarely is the state considered as an actor in its own right or seen as complex, multileveled, subject to a wide range of influences, and having contradictory interests and permeable boundaries.

Although the higher education policy literature looks at matters of curriculum, usually under the rubric of quality, the epistemology employed usually takes knowledge to be either behavioristic or pragmatic. If it is behavioristic, knowledge is seen as testable, falsifiable, tentative, methodical, yet ever more comprehesive and complete. If it is pragmatic, knowledge is seen as contributing to individual mobility, or as the solution to concrete social and technical problems.[26] Knowledge is not thought to be value laden or to incorporate ideologies which may influence the way questions about access and equity, or excellence and autonomy are framed and answered.

Business and Higher Education

After World War II, the federal government began to play a greater and greater role in supporting postsecondary education, initially by sponsoring research. Federal sponsorship of research increased every year from 1955 to 1968.[27] Corporations, which had been major research sponsors through industrial laboratories in the 1930s and 1940s, ceased to play a significant role.[28] However, federal support of research stopped growing dramatically in the mid-1960s, and, for a few years in the early 1970s, monies actually decreased slightly.[29] At this point, the policy literature began to speak about reinvigorating the partnership between business and higher education. This literature has been developed primarily by administrators and academics at elite research universities and by national science organizations ranging from the National Science Foundation to the National Academy of Engineering. In the main, it tries to discover ways to increase the flow of corporate dollars to science and to contain the federal government's investment in science by privatizing at least some aspects of basic research. This policy literature advocates a closer alliance between the two sectors and addresses (1) business as a possible sponsor of research, (2) science as a stimulus for economic development, and (3) universities as entrepreneurial actors in their own right.

The first of these literatures—the one which sees business as a possible sponsor of research—was prompted by the Nixon economic stimulus. After defense spending was sharply curtailed by the Mansfield Amendment and the economy moved into recession, Nixon put compensatory monies into the National Science Foundation (NSF) for the development of business-university research partnerships.[30] At approximately the same time—the mid-1970s—a number of startlingly large contracts between research universities and corporations were signed. A policy literature, initially sponsored primarily by the federal government and academic associations concerned with research, began to emerge that focused on corporations as research patrons able to provide a critical edge in terms of funding. Examples are provided by the National Science Board's *University-Industry Research Relationships,* Thomas W. Langfitt's *Partners in the Research Enterprise,* and Bernard Reams' *University-Industry Research Partnerships.*[31]

The problems with this policy literature are several. First, corporate leaders consistently take the position that business is unable to provide any significant amount of increased resources to higher education. As a recent Business-Higher Education Forum report puts it, "higher education institutions must recognize that business corporations, unlike foundations, are not created for the purpose of making contributions to education and other non-profit sectors of society."[32] After the initial wave of contracts dealing with biotechnology, the number of partnerships involving significant sums of money and long-range fundamental research diminished.

A second and related policy literature focuses on research universities' role in economic development. This literature is most developed at the level of the several states, and tends to center on regional economic difficulties. By and large, it looks to the role that small, high-technology firms play in job creation and in technological innovation. Examples are *The High Technology Connection and Entrepreneurial Science,* as well as reports from the U.S. Office of Technology Assessment, Department of Commerce, and the host of reports issued by state research corporations and partnerships.[33]

This literature also has problems. There is very little empirical evidence of direct linkages between university research and industrial innovation.[34] Although the small firms that pioneer high technology jobs may generate more new positions than any other sector, these firms are historically the most unstable and therefore the most vulnerable to job loss.[35] Moreover, not all postsecondary institutions are able to participate in regional economic development. Those most likely to play a significant part are research universities.[36] Finally, it is not at all clear that research universities will be able to deliver what they are promising: new products and processes that will contribute to regional prosperity.[37]

Another closely related literature speaks to the role universities can play as economic actors in their own right. Although this is probably the least well developed of the literatures on the relationship between corporation and campus, it is nonetheless a distinct strand. The university as joint partner in economic enterprises, as equity shareholder in its own faculty's firms, and as venture capitalist are central topics. The presidents of research universities with successful relations with industry and strong patent records are most articulate spokespersons for these ideas. For example, Derek Bok and Donald Kennedy have written at length on the interest of research universities in these matters.[38]

The difficulties with this position are fairly obvious. There is a strong possibility of conflict of interest between individual faculty members and the university.[39] There is also a possibility of competition with private sector economic enterprise, against which there is often legislation at the state level.

In general, the mainstream literature on business and higher education rests on a series of somewhat questionable assumptions. The literature assumes that the benefits of discovery will be evenly distributed, falling equally across a region's population, an assumption that is not well founded. It assumes that the several states will contribute an increasing amount to funding research and that the federal government will continue to play a significant role in underwriting research, yet the role of the state is not discussed. Finally, the literature assumes that the university will be able to discover and deliver profitable technology.

There is a long literature that is critical of corporate and campus relations. It began to develop around the turn of the century, as corporations and universities started working together on research and training. This literature ranges from Veblen's *The Higher Learning*, in essence a technocratic critique, to Social Gospel objections to foundation gifts to higher education as "tainted money."[40] The most enduring strand in the critical literature sees corporate capitalism as inherently exploitative and thus antithetical to the values of the university. Scholarship in this vein began in the Progressive Era, continued through the 1920s with work such as Upton Sinclair's *The Goosestep*, in the 1930s and 1940s with Ernest Victor Hollis's *Philanthropic Foundations and Higher Education* and Hubert Park Beck's *Men Who Control Our Universities*, and in the 1960s and early 1970s with James Ridgeway's *The Closed Corporation*, Irving Louis Horowitz's *The Rise and Fall of Project Camelot*, Bowles and Gintis's *Schooling in Capitalist America*, and David N. Smith's *Who Rules the Universities*.[41]

In the 1970s and 1980s, the critical tradition, with some exceptions, has not flourished.[42] Part of the explanation for the recent lack of development of this tradition may be found in the model on which it relies. The critical tradition has seen the interests of corporate capital as central to the ordering of higher education. Although these studies point out some of the ways in which class influences higher learning, they tend to be structural critiques and are often overly deterministic. They frequently see corporate capital as a unified entity and therefore look neither for conflict among elites nor for principles of cohesion between elites and other factions. Little attention is paid to the state generally, let alone to the various branches and levels of government. Finally, not much concern is expressed about the way in which academics as an occupational group fit into the class structure.[43]

Science Policy Literature

The science policy literature is primarily concerned with research and development, and therefore focuses almost exclusively on graduate schools at research universities. There are three main branches in the science policy literature. First there is the literature which looks at the determinants of discovery, status, and prestige. Second, there is the literature concerned with the priorities and consequences of science policy. Third, there is a neo-Marxian literature which focuses on whom science policy benefits.

The discovery, status, and prestige literature is a durable strand. In the main it argues for increased funding based on linkage between federal resources, graduate science programs, quality, scientific discovery, and international competitiveness, the last-mentioned including development of technology, as well as economic and national security. The literature is grounded theoretically in an empirical positivist sociology of science represented by scholars such as Warren Hagstrom, Diane Crane, and the Coles.[44] In essence, this tradition asks whether the scientific reward system is meritorious or not, and finds that it is. Highly rated scientific programs are seen as producing scholars who make the most notable discoveries, win the most prizes, and consistently contribute frequently cited work to their fields. When these indices and arguments are employed in science policy literature, they make a case for continued, high-level funding by the federal government of science programs at elite research universities. Prototypic examples of this work are Dael Wolfle's *Science and Public Policy,* and Bruce Smith and Joseph Karleskey's *The State of Academic Science.*[45]

There are several difficulties with this literature. Science is presented as both value free and of the highest value. The various disciplines are seen as engaged in "objective" or "hard" science. Contained within this epistemology is an implicit ideology of discovery, progress, and the possibility of technical solutions to social and economic problems. Because science is presented as value free, the assumptions behind this view are not systematically interrogated. The status and prestige literature ignores the state even while demanding more funds from it. The purposes of the several mission agencies, from which most research and development funds flow, are by and large disregarded. The state is considered mainly in its capacity as a science resource provider. In policy terms, more funding for science is almost always good policy. Niether the scientific establishment nor the mission agencies are seen as motivated by anything other than excellence, scientific competition, and progress. The social model employed is meritocracy, and questions about the way class, power, and profit might affect definitions of quality are unasked.

The second strand in the science policy literature, that which is concerned with the priorities and consequences of science, sees the articulation of science with policy and the economy as more problematic. On the one hand, it raises questions as to whether particular programs—for example, the Strategic Defense Initiative—should be funded; on the other hand, it asks how science should be regulated. Examples of the literature are Kuehn and Porter's *Science, Technology and National Policy* and Goggin's *Governing Science and Technology in a Democracy.*[46]

This literature puts science in social context and takes account of the political and economic problems generated by science. Of special concern are decision-making processes. For example, this literature looks at risk management, contracting processes, and unintended consequences of policy. However, it focuses not so much on class and power as on interest groups; not on theories of the state, but on state mechanisms for accountability; not on ideology, but on reasoned argument aimed at parties interested in the decisions at hand. This literature, then, looks to a liberal, interventionist state to regulate problems— such as procurement for military science, excesses in business research and development, or unregulated corporate development of new genetic material.

Third, there is a neo-Marxian strand in the science policy literature. It is represented by authors such as David Dickson, Daniel Greenberg, Martin Kenny, Sheldon Krimsky, Dorothy Nelkin, and David Noble.[47] This literature sees science as shaped by class struggle. It concentrates on the ways in which corporate leaders articulate with the scientific establishment and the state, and by and large sees corporate interests as shaping the kinds of science that we do. Science policy is very often interpreted as business policy.

Although the neo-Marxian literature presents valuable case studies of the way in which the corporate and scientific communities interact to shape science policy, it is frequently too willing to see the corporate community as able to dictate a science agenda without encountering resistance from faculty or accommodating to academic interests. Generally, corporate leaders are seen as representatives of a very powerful ruling class, and the class interests of other fractions or subordinate classes—particularly professionals—are not systematically considered because they are not thought to be important. This literature does not incorporate much of the recent theoretical work on the state, although the scientific community receives by far the greatest part of its funding from the fifty states and the federal government. Finally, this literature deals almost exclusively with the economic aspects of the relation between the corporate and scientific communities, and tends not to pay attention to the cultural and ideological.

Almost none of the higher education policy literature, regardless of the policy questions or arenas addressed, uses an explicitly corporatist framework. As I have indicated, the policy literature is mainly cast within a pluralist tradition, although there is an enduring critical and neo-Marxian tradition as well. The lack of an explicit corporatist framework is unfortunate, since many of the current policy documents being formulated by industry-university-government groups implicitly rely on such a framework. They are corporatist in that they urge sectorial rather than democratic representation in the policy formulation process. Sectors are defined both economically and politically. The primary sectors are economically determined: the corporate community; higher education, in its capacity as training agent and technology source; and labor. Minorities and women are on the periphery, sometimes treated as special interest groups, at other times as having sectorial status. The bureaucratic state is seen as creating the infrastructure for policy formulation and implementation. Generally, the role of neither the public nor the electorate is discussed. Since the corporatist position is both a theoretical construct and a blueprint for action, it deserves greater consideration by policy scholars.

This review of the literature indicates that inquiry about the policy process needs to be reformulated. A reformulation is in order for several reasons. First, the higher education literature continues to be shaped by questions initially asked in the expansive period of the 1960s. Current operating support, entitlements, and infrastructure expectations are all measured against the "golden age," when the base of higher education was broadened and money was to be had by the academic community almost for the asking. Although high expectations may not be inappropriate, especially as a negotiating strategy, they can blind us to significant changes occurring in higher-education policy arenas, changes which might call for strategies more sophisticated than petulant demands to reestablish the status quo ante 1968.

Second, the higher education policy literature is both fragmented and focused on incompletely defined policy arenas. On the one hand, the policy literature is preoccupied with issues of access and equity or cost, quality and autonomy at the undergraduate level. On the other, the science policy literature is concerned with monitoring or increasing the flow of support to graduate programs. The two policy arenas are rarely connected. An exception in this regard is John T. Wilson, *Academic Science Higher Education and the Federal Government 1950-1983*.[48] Although Wilson treats both the federal programs that touch the interests of community colleges and four year schools and those that fund science in graduate programs, he deals with them seperately, and does not explore the ways in which policy developments in one sector may affect the other. Like other scholars, Wilson treats both these arenas primarily in terms of actors who are at center stage only in the later stages of policy formation. Treatment of access, equity and cost focuses on legislative developments, at which point issues have already been framed. Treatment of science policy concentrates on the role the academic science community plays in setting resource levels for fundamental research, by and large ignoring the mission agencies that supply the greatest amount of money to academic science, thereby shaping science agendas through the allocation process. Finally, Wilson, like other scholars does not give much attention to patterns of interaction between higher education policy actors and policy formation groups in the wider political economy.

The policy literature has generally overlooked changes in the political economy with which some leaders in the corporate and university sectors have already begun to deal. These changes are occurring in response to increased global economic and intellectual competition, and provide both constraints and opportunites for higher education policy actors intent on shaping new alliances that will enable their institutions and constituencies to be well positioned in the future. I will now turn to the Business–Higher Education Forum in an effort to show concretely the importance of these new alliances for the policy process as a whole.

The Business-Higher Education Forum

The Business-Higher Education Forum is both part of a long tradition of close relations between U.S. corporations and universities and a new departure. CEOs have long seen universities as central to corporate interests. In the words of Alfred P. Sloan, Jr., former chairman of General Motors:

> When the annals of our time are recorded, it will most likely be found that the two greatest contributions of our time have been the U.S. university and the U.S. corporation: both mighty forces, both uniquely American. If these two forces can go forward together in understanding and cooperation, there is perhaps no problem beyond their joint power for resolution. If, however, they choose to go their separate ways, there is no solution of any problem affecting either that is likely to be long lasting.[49]

However, current relationships between corporations and research universities differ from preceding ones in several ways. For the first time since World War II, the U.S. position in global markets has changed; we can no longer assume easy economic dominance of nonsocialist countries. The U.S. cannot compete with the lower labor costs of newly industrializing nations. The solution usually proposed for U.S. economic difficulties is greater emphasis on high technology, closely controlled by tougher intellectual property laws. This situation creates greater pressure for corporations and universities to engage in technology transfer and other science- and technology-intensive activities, to better enable the U.S. to regain its competitive edge. University administrators, along with many scientists, have become willing partners in "advanced applied" research, working to incorporate basic science into products and processes with commercial potential. Because of changes in intellectual property laws as well as customs and traditions surrounding the use of university-generated knowledge, for the first time higher education institutions are able to reap directly pecuniary benefits from discoveries made by the faculty. Patent royalties and equity positions in high-technology companies hold out the hope of a steady stream of resources to universities, resources for which the academy is beholden neither to private donors nor the government. Moreover, the government has begun to subsidize commercially oriented research. The federal government has broken a long standing pattern of research and development funding that focused primarily on defense or health, and is now funding university-industry partnerships as well as writing legislation to facilitate technology transfer. The several states have also greatly increased the amount of money committed to technology initiatives, spending $700 million in 1986, a significant part of which is available to universities.[50]

Although some of these changes in the relations between business and higher education stem from changes in global economic conditions with which the university was relatively uninvolved, others, such as legislation favoring technology transfer and supportive of university ownership of intellectual property, as well as federal and state funding for commercial research and development that incorporates basic science, are due at least in part to the activity of new organizations that represent the shared concerns of the corporate and academic communities in the higher education policy process. The Business Higher Education Forum, begun in 1978, was one of the first of this genre of organizations. According to its literature, the Forum is deeply concerned with influencing policy formation:

> The Business–Higher Education Forum was created for the express purpose of promoting discourse and acting on issues shared jointly by American business and the nation's higher education institutions. The Forum's major objective is to provide an opportunity for interchange among its members on matters pertaining to their respective interests and, hence, to the common aspirations and needs of the business and academic communities.[51]

The Forum was organized by: "far-sighted academic and corporate chief executives [whose] goals were to address issues of mutual concern to corporate and post secondary education communities and to pool resources for benefit not only of the two sectors, but for society as a whole."[52] The initiative for the Forum was taken by Dr. Jack Peltason, the president of the ACE. ACE is one of the longest-lived policy organizations in higher education. It was founded as a vehicle through which universities could respond to the crisis of World War I, and continued thereafter as an umbrella organization under which all of the higher education associations found shelter as they emerged. Although

ACE represents all of postsecondary education, it is commonly thought to be dominated by research universities, particularly those elite institutions represented by the AAU. The Forum, then, is a fresh organizational form put forward by an established policy organization to deal with changing parameters in the higher education policy arena.

Forum membership is composed of approximately 40 CEOs of large corporations and 40 presidents of postsecondary institutions. Of the presidents, 56 percent are members of the AAU. The AAU is an exclusive organization of roughly 50 highly ranked research universities. It offers membership only to those applicants approved by a two-thirds vote of the membership. The AAU is even more venerable than ACE. It was started at the turn of the century as a forum in which presidents and deans of newly established graduate schools could meet to share their common concerns. The initial membership was quite limited, carefully balanced between private and public universities. Membership has grown very slowly, and continues to be highly selective. For many years, the AAU functioned as a presidents' club. It met annually, but no minutes were kept and no permanent office, let alone staff, was maintained. In the late 1950s, the AAU opened a Washington office and became more active in the policy process. In the 1970s, as federal funding became tighter, the AAU became more aggressive. Currently, it, like ACE, is involved in a number of organizations developing industry-university-government partnerships.

Another 20 percent of college presidents in the Business–Higher Education Forum represent highly selective liberal arts colleges and universities.[53] If these are included with the heads of AAU universities, approximately three-quarters of the presidents are drawn from elite institutions. The university presidents participating in the Forum, then, are established national policy actors able to bring their skills to bear in creating a climate favorable for higher education in the crisis engendered by America's faltering in the world economic sweepstakes.

The approximately 40 corporate leaders by and large represent U.S.-based multinationals. Ninety percent of the CEOs are heads of corporations ranked in the top 200 of the Fortune 500. The large majority are interested in higher education, often having served on a university board of trustees, and have wide policy experience. Many are members of either the Committee for Economic Development or the Business Roundtable. The Committee for Economic Development was established in the early 1940s to shape policy in the postwar era. Initially, it was composed of two hundred corporate leaders who met annually. The Committee has been active in national policy formation for almost half a century. The Roundtable was created in 1973 to press more aggressively for business interests in the wake of the social unrest that accompanied the Vietnam War. It became a formidable lobbying organization for the business community, actually using CEOs to present Roundtable positions forcefully on Capitol Hill. Like the university presidents, the CEOs are veterans of the policy arena, ready to make common cause with their academic counterparts to promote policy that will enhance shared goals.

The Forum is a self-consciously elite organization. Membership is by invitation only. The Forum meets semiannually. The reports it produces are written by the members themselves. As Forum publicity says, it has "always been a principals-only organization," working on carefully selected issues in an atmosphere that "breeds both collegiality and consensus."[54]

Although the Forum has a strong interest in the development of high technology, which is the special province of research universities, it is also concerned with training at all levels of the postsecondary system, with the growth of scientific and technical competence throughout the educational system, and with educational excellence generally. The Forum's attention is not focused exclusively on science and research universities, but occasionally includes all of postsecondary education. As such, it is an organization that begins to bridge the gap between policy initiatives that focus primarily on undergraduate education and those that take up only issues of science policy.

Because the Forum is primarily interested in policy formation, measuring its impact is difficult. It is possible, however, to get a sense of the Forum's place in fashioning entrepreneurial conditions that promote new alliances between business and higher education by noting a few of the policy positions strongly promoted by the Forum: research tax credits, technology transfer legislation, university and corporate ownership of patents developed in the course of conducting federally sponsored research, lifelong learning for professionals, space-grant universities, industry-university-government production partnerships, and knowledge-pooling agreements among enterprises and universities working on similar problems. Although the Forum is one of a number of organizations concerned with creating close relations between business and higher education, it is national in scope, has brought together representatives of powerful organizations in the several sectors, is closely connected to established policy-making organizations, and is very active. An examination of the Forum should enable us to learn something of the new directions that established actors in the policy formation process are taking.

The Higher Learning and High Technology

I come to this study of higher education policy from a Marxian tradition. By this I mean that I use a historical perspective, am a materialist, insofar as I see resources and power as central to determining the allocation of social goods and services, look initially to class and the relationship of class actors to the forces of production when trying to account for policy outcomes, and find technical and ideological knowledge to be inextricably entwined. I use this as a point from which to initiate inquiry, not as a foregone conclusion. In other words, I see the relationship of higher education to the political economy as problematic, not necessarily economically determined, a question which must be constantly theorized and on which data must be brought to bear.

Although I find recent neo-Marxian theory rich and intellectually stimulating, especially with regard to its treatment of the state and ideology, I try to remain cognizant of the limitations of this theoretical tradition. I do this by trying to raise consciously the theoretical issues at stake in each chapter of this book, and then considering the data I have brought to bear on these issues in terms of the explanations offered by pluralism, neo-Marxism, and corporatism. I try not to prejudice my answers, and hope that sharing my theoretical predisposition with readers will keep me from manipulating my own text too obviously.

In the following, or second chapter, I look at the history of higher-education policy in the post–World War II period in an attempt to understand the changing context in which the Forum works. I focus on changing structural conditions in the broader political economy, the conflict between educational haves and have-nots over state resources and within state agencies, and the realignment of organizations and constituencies interested in higher education policy formation in response to these conflicts and changes. The policy-making configurations that emerge are reviewed in terms of the theoretical positions discussed earlier, with attention given to the strengths and weaknesses of each theory in understanding policy dynamics in the postwar period.

Chapter 3 looks at the social-class location of Business–Higher

Education Forum members. The focus of inquiry is on their class location, the similarities and differences between corporate and campus leaders, and their positions in broad corporate and university policy-making structures. The data base was compiled from a wide variety of biographical dictionaries and encyclopedias. The method used is power structure research. The theoretical issue in regard to the social composition of the Forum is the position of these leaders in the class structure. Are they members of the upper class, or do they display a wide variety of class backgrounds? If they are not drawn from an upper class or ruling class, is there any obvious way in which they are connected to such a class? Do CEOs and university presidents have different positions in the class structure?

Chapter 4 looks at policy positions taken by university leaders prior to their joining the Forum. This inquiry allows us later to make a judgement as to whether the Forum represents a continuation of previous efforts, a break with past policy positions, or a partial redefinition of their work. The AAU was selected as the organization most representative of university presidents' past policy efforts. The data base was constructed from fifteen years of AAU presidents' testimony before Congress on behalf of their institutions or the Association. The method employed is content analysis of presidential testimony. The theoretical issue of interest is the way in which presidents are able to gain access to the policy process. To what degree are leaders class-conscious social actors? If they are not informed by class consciousness, what is the basis of their social cohesion? How do they order their policy agendas and explain change and progress: which has primacy, the economy or the institutions which create ideas, technology, and ideology?

Chapter 5 looks at the characteristics of the corporations that are members of the forum, at their involvement in higher education in terms of support for research, scholarship, and donations, and at the positions CEOs take with regard to policy in the popular press. Several data sets were analyzed—ranging from corporate rankings, as given by *Fortune*, to corporate statistics, as compiled in *Moody's*—using a variety of techniques, ranging from economic inventories to content analysis. The theoretical issues here parallel those raised for university presidents in the preceding chapter but focus on the economy rather than the academy.

Chapters 6 and 7 are content analyses of six documents issued by the Forum over a four-year period. Because the majority of the recommendations for action stemming from the Forum's reports are focused on government, it is possible to address a pivotal theoretical issue: the relative autonomy of the state vis-à-vis a politically active segment of corporate and university leadership. Given the continuous concern of the Forum with the importance of developing a broad societal consensus on social and economic priorities, it also becomes possible to explore theoretical questions about ideological hegemony in a concrete situation.

Chapter 8 reflects on the implications of the previous chapters, giving particular attention to questions about class, state, and ideology in the higher education policy process. The theoretical positions explored earlier—pluralism, neo-Marxism, and corporatism—are reconsidered in light of the data and arguments made in the course of the book. From this analysis, new problems are formulated for scholars of higher education policy.

In conclusion, I want to say a few words about the metes and bounds of this study. *The Higher Learning and High Technology* is primarily a historical account, rooted in document analysis. Although it draws from a wide variety of documents, it suffers from all the problems that plague such work. The internal politics that figure in the creation of the documents used are unknown. The conflicts and compromises involved in writing these reports and the degree to which they represent the principal actors involved are not clear. Even authorship is uncertain. Similarly, the degree to which organizations such as the Forum actually influence the policy-making process is not explored. And no matter how powerful the organization and how sophisticated the position papers developed, resistance to and negotiation over proffered policy by a wide variety of groups is perhaps the only certain response to organizations such as the Forum. The policy positions of the Forum are only one counter in the elaborate and ongoing policy game. Despite these very real limitations, I think what the Forum is doing is worth our attention, because the Forum brings together central actors and organizations in the policy process who are intent on forging a new consensus on higher education policy issues for the twenty-first century. Hopefully, other work will answer the questions precluded in this study by time, choice of subject, framework, and methodology.

CHAPTER 2

Policy Issues Past and Present

In this chapter, I provide the historical context for the emergence of the Business-Higher Education Forum and similar organizations. I begin by outlining changes in global economic conditions with which U.S. corporations currently contend, emphasizing those points of concern to both the business and the academic communities. Generally, I argue that increased competition in world markets leads to increased competition for domestic resources on the part of both the corporate and academic communities, prompting business and university leaders to take more aggressive positions, especially in the political arena, as they attempt to maintain their respective domains.

Next, I identify higher education policy questions that have preoccupied educators and legislators since World War II. I see these questions or issues as ones of access, career preparation, the development of national research goals for the sciences, and patterns of resource allocation among these several functions. I present a broad overview of how these issues have been handled since World War II so that we can better understand how economic constraints are reshaping these issues and why business and academic leaders are developing variations on their patterns of policy intervention.

As I speak to major policy issues, I also address the political and economic constituencies that are built around them. In other words, I am concerned with class and interest groups, as well as with coalitions formed around race, gender, and ethnicity, and the part they play in shaping higher education policy. I am particularly concerned with the predictability and stability of responses on the part of the various constituencies.

I then outline the concrete legislative agenda being developed by the Business-Higher Education Forum and try to show how it emerges in response to the social conditions, policy issues, and political climate that I have described. I indicate the ways in which the Forum's initiatives are outgrowths of past policy and where they diverge. I conclude by reviewing the historical account of higher-education policy presented here through the several theoretical lenses I have discussed: pluralism, neo-Marxism, and corporatism.

Changing Economic Conditions

Among the more significant changes in the political economy in the past two decades are the development of a global economy and the beginnings of a global division of labor, the falling rate of profit, and the growth of high technology as a dynamic profit sector, along with service as a significant one. The consequences of these changes for the corporate community are fairly obvious. The development of a global economy has increased the number of players trying to win large shares of world markets, with a consequent heightening of competition worldwide. Developing countries have an advantage in terms of low-cost labor, which, together with tax structures, contributes to the movement of American manufacturing offshore, a situation that further weakens domestic production. The falling rate of profit has intensified competition yet again. In response, corporations are pushing to increase labor productivity, often through technology that emphasizes ever-increasing automation, especially computers and robotics. High technology is also emphasized in its own right. It is seen as a way of enhancing our economic position without making major changes in production patterns and concomitant changes in the social order. Generally, there seems to be a widespread belief that high technology, properly exploited, will help us regain our position in global markets.

There are a number of points at which changes in economic conditions touch higher education, highlighting areas in which corporate and academic interests converge. The development of a global economy and the beginnings of a global division of labor intensify competition at all levels and in all sectors.[1] Economic rivalry is mirrored by conflict over scientific rewards, both at home and abroad. Competition takes place within and between nationally based universities, and is focused on winning large research contracts with industrial sponsors and on the discovery of patentable and profitable products and processes. Increased scientific competition, especially entrepreneurial science, heightens existing tensions within and between universities over the free flow of information, nationally and internationally, the free movement of students and professors, the ethical limits of private exploitation of the knowledge developed in public universities, and the autonomy of the university from both state and corporate sectors.[2]

A falling rate of profit increases the likelihood that national and state capital will be spent to meet business rather than social needs.[3] Public monies which are used for social expenditures will probably be much more closely examined. For example, higher education costs are currently evaluated in terms of human capital theory, which treats most curricula equally.[4] In the future, costing theories will probably be developed to account for specific programs and their contribution to private-sector profitability. All in all, fiscal crises, exemplified by falling rates of profit, probably mean that greater care will be given to the definition of various segments and curricula in higher education, with only those most likely to contribute to private-sector growth receiving greatly increased investment.[5] Such an area of investment is apt to be high technology. High technology is currently seen by business, higher education, and political leaders as a means of dealing with the fiscal crisis caused by falling rates of profit.[6] Basic research is increasingly presented as the key to unlocking new and profitable technologies. Indeed, basic research in graduate universities is in the process of being redefined as entrepreneurial science.[7] Entrepreneurial policy includes the development of university-industry partnerships around shared research agendas, the fostering of technology transfer between universities and the private sector, the growth of university-sponsored incubators for science-related, high-risk businesses, and the fostering of companies that "spin-off" from university-patented research, providing the university with royalty income. Both corporations and universities benefit from entrepreneurial science.

These general trends—the growth of high technology as a dynamic profit sector, a falling rate of profit, and increased competition in international markets—have prodded leaders in the corporate and academic communities to organize more effectively. In the 1970s, established organizations representing corporations began to address these trends, and new groups were created specifically to deal with them. Established corporate organizations which have made international competitiveness central to their policy agendas are the Committee for Economic Development, the Conference Board, and the American Business Conference. New organizations include the Business Roundtable as well as think tanks such as the American Enterprise Institute. Higher education associations reveal a similar pattern. Established organizations, exemplified by the National Science Foundation, the National Academy of Engineering, the American Association for the Advancement of Science, the Defense Department, the Association of American Universities, and the American Council on Education, have begun to address issues of international competitiveness and entrepreneurial science. New organizations that reflect higher education's interests in these areas are the University-Industry-Government Roundtable, the Department of Defense-Higher Education Forum, opinion outlets such as *Industry and Higher Education,* and, of course, the Business-Higher Education Forum.

These organizations and associations articulate the interests of the corporate sector and entrepreneurial science to the broader community in a conscious effort to shape the growth of high technology in a manner that contributes to business prosperity. In so doing, these organizations have to influence or accommodate the political forces that have coalesced around the major policy issues that have emerged in the postwar period. These issues are access, goal setting for federal science policy, and allocation of resources.

Access

Access to higher education has been a central issue in the twentieth century. Higher education and the professions to which advanced degrees lead became the last frontier, the arena for proving individual prowess and reaping social rewards. The meritocratic ideology embedded in higher education serves much the same function as social Darwinist ideology did in the late nineteenth century: it justifies individual rewards without attending to the social structures that support or inhibit individual effort. According to meritocratic ideology, those with the greatest intellectual capacity enter into strenuous competition with their fellows, in which the winner is named the fittest, or the most meritorious, and, claiming a credential rather than the means of production, invariably goes on to a long and prosperous life. All the elements of the social Darwinist ideology are present: native ability, individualism, hard work, and severe testing through strenuous competition, which selects the most able. Thus, access becomes a central social issue for all individuals and groups with upward aspirations.

What is often absent from the discussions about social Darwinism and meritocracy are considerations of class and resource monopolies. As industrial competition was subordinated to monopolistic organization in the economic sector to preserve class privilege, so educational competition was contained by a certification system designed by emerging professionals as a means to provide the middle class with predictable social and economic rewards.[8] Certification turned on skills and attitudes possessed in abundance by the middle class—cultural literacy, numeracy, perseverance, self-confidence, appropriate assertiveness, and socially acceptable manners—and not found as frequently among immigrants, the working class, or the working poor. Professional certification became a way of organizing knowledge that reinforced prevailing patterns of privilege. Indeed, at the end of World War II, the nation's leading universities by and large still served white, Anglo-Saxon, Protestant males.[9]

Initially, demands by those excluded from higher education—working class youth, minorities, women—were for the most part not directed toward the redistribution of educational goods and services, but toward obtaining economic benefits. In the wake of World War II, industrial unionism seemed as likely a vehicle as higher education for the aspirations of American workers. However, the radical energy of working-class youth was contained by high wages in the enterprises most affected by militant trade unionism, and the ambition of veterans with aspriations for upward mobility was channeled into higher education. In terms of higher education, the national policy response, at least for white males seeking material rewards and status, was to substitute educational for economic goods, which led to increased access to higher education, precipitating the great expansion of the 1950s and 1960s.

Even though the expanded academy continued to block access for a number of groups through certification systems that raised formidable barriers in terms of preparation, testing, culture and class identification, and even overt discrimination, the rhetoric and ideals of meritocracy, which were central to maintaining higher education as the appropriate vehicle for social mobility, still provided levers for change. For a credible meritocracy, equality of educational opportunity was necessary, and higher education had to keep growing. After the GI Bill, which offered an avenue for upward mobility to white men, came the 1965 Education Act, which finally made it possible for blacks to realize some of the gains made in the 1954 Brown decision.[10] Title VII and IX of this same act, together with Executive Order Number Four in 1972, gave women too an equal right to higher education. P.L. 503 provided physical access to higher education for the handicapped.[11]

The offspring of working-class families, women, blacks, and other minorities genuinely benefited from this expansion. They significantly raised their completion rate for the B.A. and higher degrees, and frequently moved into occupations and careers quite different from those of their parents. Initially, this expansion of higher edcuation was not threatening to the established order. The private sector was asked for very little. Growth was funded with public dollars in a period when taxes on corporations were rapidly decreasing and loopholes for the wealthy were increasing.[12] In effect, more accessible higher education was paid for by taxes on the working class, especially the prosperous blue-collar sector.

However, previously excluded groups often received a differential rate of return on their educational credentials. Despite their possession of similar credentials, white males generally continued to earn more. The 1982 unemployment rate for blacks in science and engineering is more than twice that for whites, and on average blacks earn lower salaries.[13] Women scientists and engineers too have an unemployment rate about twice that for men and an annual salary rate that continues to average about 80 percent of that for men.[14]

Even as the promise of social mobility through advanced education was not fully realized by all participants with similar credentials, the cost of maintaining higher education rose. To make access more equitable and merit more plausible, a greater and greater share of public resources had to be allocated to higher education. In the 1960s and 1970s, federal and state legislation authorized funds for new institutions, dormitory construction, student loans and grants, and broader programs ranging from special services for underprepared students to women's athletics. The increasing cost to the federal and state governments is dramatically revealed by figures for the income of institutions of higher education. In 1950, the federal government supplied approximately $.50 billion dollars to all U.S. postsecondary institutions; in 1976, it supplied $5.41 billion. In 1950, the several state governments also provided approximately $.50 billion dollars; in 1976, they contributed $12.26 billion.[15]

As costs soared, many business groups, often aided by major foundations, began to move for containment. For example, the Committee for Economic Development in the 1970s moved to increase tuition to the point where students would pay about half the real cost of their education. Under the Reagan administration, families were asked to bear an increasing share of their children's college costs. As the deficit continued to mount and the U.S. continued to lose ground in global markets, social programs—remedial, tutoring, skills development— that contribute to the increased enrollment of previously excluded groups were cut, grant and loan programs came under increasing scrutiny, and low-cost community colleges were emphasized as an institutional answer to demands for improved access and equal educational opportunity.

In response to these changes, previously excluded groups—blacks, women, and other minorities—are trying to understand why they have not realized the same rewards from higher education as have the academy's more traditional constituencies. At the same time, they are asking that educational opportunity continue to be expanded. Middle-class organizations are trying to protect and improve subsidies that have allowed them to educate their children easily: state funding of quality public institutions, scholarships, grant, and especially loan programs.

Organizations such as the Business-Higher Education Forum have not spoken against the expansion of higher education. Indeed, they sometimes address the importance of maintaining a national commitment to educational opportunity for blacks, women, and other minorities. However, the Forum's economic priorities and plans for allocation of resources contradict and, in effect, undermine its concern with expanding access and equal educational opportunity. The Forum's legislative agenda, discussed later in this chapter, emphasizes putting public resources in programs that ever more directly subsidize private industry, thereby inadvertently reducing the monies available for social and educational programs. It calls for the reduction of regulatory programs in the name of unfettering business energies, which means that affirmative action programs are undercut. It asks that educational resources be concentrated in programs designed to train scientists and engineers, areas which have traditionally not been hospitable to blacks, women, and other minorities.

Thus far, the discussion has centered on the vertical expansion of higher education, or the ways in which opportunities for access and upward mobility for individuals and particular social groups have increased. As vertical access increased, horizontal, or geographical and institutional access, necessarily broadened. Horizontal expansion presented its own set of problems and possibilities with regard to the issue of access.

At the end of World War II, the nation's leading universities were in the main located in the Northeast, Midwest, and Far West. These several dozen universities were, by and large, members of the self-consciously elite Association of American Universities. They ranked highest in scholarly rating schemes, received by far the lion's share of federal research and development monies, and claimed to attract the brightest students.[16] However, popular pressure for increased access fueled the growth of new institutions as well as the upgrading of old ones in areas previously unserved by nationally ranked universities. The upgrading of state universities in underdeveloped areas of the country was further stimulated by rapid economic growth, which in turn fostered higher education growth. Higher education growth gave rise to increased research capacity, and, as the National Science Foundation notes, in the 1970s, research and development monies had "much higher rates of growth in the sunbelt states of the South and the West than in the more northerly regions."[17] So too, the American Council on Education's periodic quality rankings revealed upgraded state research universities as making progressively better showings.[18]

The growing ascendancy of public universities in rising states seemed tied to strategies of capital development over which established universities in declining states had little control. Cheap labor was a key factor in the resource shifts which so drastically affect university fortunes.[19] In the decade 1970-1980, states without unions had the greatest growth in new jobs, the greatest growth in higher education support, and a substantial increase in federal research and development monies for universities. Contrarily, states with laws that protect unions lost jobs, had a marked decline in state support for higher education, and were somewhat less able to attract federal dollars for university research.[20]

These resource shifts helped broaden the geographical distribution of higher education, decentralized educational opportunity, and made access easier for previously unserved groups, especially those unable to afford the costs of going far away to college. However, the broadening geographical distribution was not an unmitigated victory in the struggle to democratize higher education. Again, those least able to bear the cost paid the greatest price. Plant closings, union busting, high unemployment, and economic insecurity became endemic in once strong union states. Higher-education institutions in these states were plagued by fiscal crisis, which resulted in reduced budgets, and the loss of outstanding faculty and their research contracts. Moreover, the gains made by previously underdeveloped states were not permanent.[21] As workers organized and taxes increased in these states, manufacturing plants moved offshore. Rapidly expanded postsecondary systems were trimmed through retrenchment, and the future for the higher learning became uncertain.[22]

Although a large number of state universities were significantly upgraded in the 1960s and 1970s, overall shifts in national prestige and resource hierarchies were not great. Instead of the twenty highest federal research and development recipients accounting for over one-half of all federal research funds, they now account for approximately 40 percent.[23] Although many departments at upgraded universities score higher in American Council on Education ratings of graduate programs, few departments at established universities rank lower. Over the course of time, a very few established public universities—Columbia, Cornell, Johns Hopkins—have slipped from composite ratings of the "top ten," but these have not been widely replaced by institutions from rising states.[24]

However, established public university presidents were unnerved by changes in economic and educational trends. They did not see themselves in a position to cope with growth patterns that threatened to pass them by. Representatives of some private universities too were threatened by deteriorating state and national economic conditions. Status and resource anxiety probably explain the readiness of the many administrators of prestigious, long-established institutions, in the private and public sectors, to join forces with organizations such as the Forum. These corporate policy formation groups promise resources to established universities in those areas which have traditionally served to maintain or enhance their status and prestige: basic science, and research and development. To claim these resources, universities have to redefine basic science as entrepreneurial science, readily translatable to commercial products and processes, and research and development as intellectual property, tightly held by corporations and academic institutions. University administrators are ready to enter partnerships with industry based on these definitions because they are invited to share in the profits promised by entrepreneurial science.

The issue of access, then, continues to be central to higher education policy making as the century draws to a close. Corporate and higher educational policy-making groups, such as the Business-Higher Education Forum, seek public resources for private enterprise, particularly manufacturing, for product-oriented research programs, and for curricula that train persons able to lead corporations in their quest for greater shares in the global market. Whether intentionally or inadvertently, realigning these policy goals reinforces traditional hierarchies of class, status, and institutional privilege.

Groups that are new to higher education continue to see it as an arena in which to demonstrate merit and win social rewards. They advocate priorities and programs at odds with the Forum. These groups want higher education to be a right, not a privilege, with full funding and a guaranteed place for every qualified student in the curriculum of their choice. They also expect similar credentials to bring the same returns, regardless of gender or race, in terms of careers and salaries. Middle-class and professional groups too support policies at odds with the Forum. They want public subsidies for quality higher education continued, along with subsidies to particular programs, such as medicine, and the maintenance of aid programs that make higher education affordable to tenuously affluent constituencies.

The Social Uses of Knowledge: Career Preparation

Although members of the university community value knowledge for its own sake, knowledge also serves social ends. Academic knowledge generally takes the form of teaching, research, or service, and each of these have a certain social utility. Teaching is central because it leads students through an orderly sequence of courses to specific sets of credentials and particular careers. Research is usually funded because it results in discoveries that benefit society. Service is important because it brings university expertise to the wider society, concretely demonstrating the social utility of knowledge. Entrepreneurial science, especially as embodied in high technology, is a dramatic way in which the university deploys its expertise. The current debates over higher education policy turn in part on the social uses of knowledge: Who is being prepared for what careers? What kinds of social priorities order research, and whom do they benefit? What kinds of expertise are being provided to which sectors of society? Again, to understand fully the stakes at issue it is necessary to consider briefly the social utility of teaching, research, and service from World War II to the present.

In many ways, the dynamic of change surrounding career preparation is similar to that surrounding access. After World War II, groups for whom the professions had previously been unattainable were able to begin using higher education to realize professional ambitions. However, the traditional professions and most prestigious universities remained relatively impervious to expanded preparation. Instead, new groups were in the main incorporated into emerging or semi-professions, or employed in professional capacities in the public sector. At the close of World War II, teaching continued to prepare elite students for leadership and elite and upper-middle-class male students for the traditional professions: medicine, law, management, and engineering. When higher education began to expand after World War II, the elite professions did not expand as quickly or as fully. The traditional professions had established nearly perfect credentialing monopolies and were able to control access closely, keeping supply tight enough to ensure a strong demand for services.[25] Although some working-class people, blacks, women, and other minorities were incorporated in the traditional professions located in the private sector, the majority were absorbed into emerging specialties, such as computer and health sciences, or into careers in the public sector. In the period 1950–1970, the greatest growth in employment was in the state sector, where there was an increase from 21 percent to 33 percent in services and governmental work, "especially in the military, welfare and educational bureaucracies."[26] The number of persons engaged in professional and technical occupations almost doubled, from 7.5 percent of the work-force in 1940 to 14.2 percent in 1970. Although government sector work often represented real mobility for first-generation college students, careers in new specialties and the state sector were very often not as lucrative and did not have the same status as the traditional professions housed in the private sector.[27]

The expansion of postsecondary education itself clearly reflects the trend toward differential career growth that was taking place in the wider society. Public-sector higher education expanded much more quickly than private. Fifty percent of students were in public institutions in 1950, and by 1976 the figure was 78.4 percent, with a corresponding increase in public-sector faculty employment.[28] The vast majority of these state institutions were not elite research universities. Women who made careers as professors were much more likely to be located at rapidly growing public institutions, especially at two- and four-year colleges, than at elite private institutions. Although new faculty positions provided comfortable livings, they offered neither the status, remuneration, nor career opportunities that came with jobs at more prestigious institutions.[29]

The expansion of higher education stimulated the growth of agencies that served it. State higher education agencies grew to coordinate growth, as did quasi-public organizations, such as accrediting associations and national organizations—for example, the American Association of Community and Junior Colleges and the National Association of State Universities and Land Grant Colleges—representing different segments of the postsecondary community.[30] The staffs of the several federal agencies that fund and monitor higher education also grew rapidly: for example, the Department of Health, Education and Welfare, the National Institute of Education, and the several organizations concerned with civil rights and affirmative action.[31] Positions in the bureaucracies and organizations are often filled by persons in emerging specialties, including disproportionate numbers of women and minorities. These specialties—and higher-education agencies provide an example—often have unclear career ladders and uncertain reward structures.[32]

New specialties and the expansion of career opportunities in the public sector were not the only mechanisms protecting the traditional professions. The "semiprofessions" also grew. In the main, they absorbed growing numbers of women who prepared for state sector careers. Schools of education, social work, nursing, and librarianship grew very quickly. In an effort to emulate the traditional professions and to share in their social rewards, these schools greatly increased standards and requirements and lengthened the course of study.[33] However, efforts to professionalize were only partially successful. The logic of state expansion, not the new professions, controlled the market. The state as well as the private sector depended on relatively modest rewards for women if costs were to be contained and the existing structure of privilege preserved.

In sum, the great postwar expansion of higher education allowed some working-class people, women, and other minorites to have careers in the traditional professions, but more often than not outside the prestigious and lucrative realm of private-sector practice and the higher levels of management. Instead, many members of these groups were drawn into public sector professional and technical work. However, even in the public sector the careers of new entrants to higher education were often differentiated by class, gender, and race, with women and minorities likely to be at the lower end of the pay and prestige hierarchy. Yet even though the collective economic gains these groups made may have been modest, the gains made by individuals were very often significant, and these, in turn, offered hope for the future to others. Expansion of the public sector was essential to expanding individual economic opportunity and collective social horizons.

Since the early 1970s, elite policy-making groups and blue-ribbon panels have advocated contraction of the public sector. These demands became more urgent during the Carter and Reagan administrations. The justification for cutbacks was the fiscal crisis of the state—the point at which tax revenues could not keep up with public expenditures.[34] Stringent retrenchment was and is presented as the only way to restore economic health.[35] Although the diagnosis of crisis is often questionable and retrenchment as a treatment untested, especially with regard to long-term consequences, cutbacks were widely introduced. The career base for representatives of groups new to professional and technical occupations was sharply curtailed, leaving them with fewer opportunites and only themselves and the invisible hand of the market economy to blame.

The crippling effects of retrenchment on the groups who benefited from public sector expansion can begin to be grasped by looking at what happened to education in the early and well-documented case provided by New York. After New York City came close to bankruptcy, the City University of New York (CUNY) began charging tuition, cut back on open enrollment programs, and fired 5,000 part-timers and 1,000 untenured full-time faculty. These professors were disproportionately young people, women, and members of minority groups who were often working in remedial and special programs.[36] A less drastic version of the same process occurred in the State University of New York (SUNY) system, where approximately 180 faculty, some of whom were tenured, were dismissed. The SUNY cuts did not result in overall reduction of employment—new hires more than made up for the firings—but the cuts did bring down the average salary in the system.[37]

The New York case illustrates several strategies for retrenchment. In the CUNY system, across-the-board cuts were instituted and new entrants to the system were most likely to lose their jobs. In the SUNY system, cuts were selective, and faculty in programs unable to muster a strong defense were likely to be fired.[38] In either case, the overall outcome was the same: less resources for the public sector, fewer career opportunities, and reduced social services. Cutting public jobs also cut public sector political strength by undercutting workplace solidarity and greatly increasing professional insecurity.

Although the decision to retrench was widely represented as stemming from economic need, views of economic necessity are often shaped by political predispositions. For example, elite groups such as New York's Commission of Independent Colleges and Universities did not ask that public spending be curtailed, but rather that it be tailored to private-sector needs. Indeed, public funding for private sector higher education in New York increased at a greater rate than in the SUNY system in the years following the initial, dramatic retrenchment. Between 1975 and 1981, state spending increased by 104 percent, with aid to private colleges and universities increasing at the rate of 65 percent and SUNY spending at a rate of only 43 percent.[39]

What is happening to higher education spending at the state level is also happening nationally, across public service programs. Efforts are being made to cut substantially public programs that distribute benefits to a wide range of citizens. These include not only programs for the poor—AFDC, food stamps, school lunch—but also programs that reach the middle class—social security, medicaid, student grants, and college loans. These cuts are justified on grounds of economic necessity; they are presented as essential to the increased free market productivity that will ultimately benefit the whole society. However, it is not clear that a free market economy will broadly redistribute wealth to the hardest working and most efficient. Indeed, true free markets probably ceased to exist by the turn of the century.

Heavy cuts in broad social benefits programs do more than release public resources for private sector use. Widespread retrenchment also erodes the resource base and political autonomy of public sector workers, creating a climate of austerity that makes political organization difficult. The university provides a case in point. The great expansion of the 1950s and 1960s created a resource base and staging grounds for a politics critical of the corporate sector and militarism, in Democratic and Republican administrations alike. The erosion of public resources has probably made the reemergence of such politics more difficult. Those areas that gave voice to the greatest criticism—the humanities, social sciences, and nontraditional subjects—have been hardest hit. Even if tenured faculty were willing to re-engage in the politics of protest, it is not certain that students would follow. The climate of austerity makes the possible cost of oppositional politics seem very high to aspiring students.

The issues surrounding the social utility of career preparation are gaining clarity. On the one hand, corporate leaders argue for public subsidy of careers that enhance the private sector on the grounds that in the long run a strong private sector is essential to a healthy national and international economy. On the other hand, previously excluded groups continue to clamor for equal educational opportunity in terms of career preparation; middle-class parents are articulate in defending the expansion of professional opportunity so that their children will be able to live in the style to which they are accustomed; social critics are beginning to speak about preparing students for a new spectrum of careers that involve some public management of an economy that would provide more direct benefits to average citizens.

The Social Uses of Knowledge: Research and Service

In the postwar era, the central social uses to which academic research has been put are health and national defense. Health research has generally been viewed as a worthwhile social investment, but the social utility of defense has been debated. I will focus on defense because its controversial history throws so strong a light on the policy dynamics central to higher education. I will also look at a new item increasingly found near the top of national research agendas: high technology research.

The policy dynamics involved in setting national research agendas for defense are fairly clear. When the United States faced crisis in its imperial domain, as was the case in World War II, the Cold War, and the Vietnam War, higher education policy makers supported expansion of the military and weaponry. The amounts of money committed to defense were so great that there was enough leeway for corporate and academic leaders to realize independent research goals even while they work with the Pentagon.[40]

This pattern was established during World War II, when representatives of the federal government, elite universities, and war industries established strong working relationships around broad defense projects.[41] Government, university, and industry joined together to develop the atomic bomb, radar, sonar, influence fuses, and other defense technology. After the war, academic research continued to serve defense needs, especially as the Cold War heated up. University presidents and spokespersons from elite research universities—Vannevar Bush from MIT, James B. Conant of Harvard, Karl Compton of MIT—supported by the executive branch of the federal government, called for liberal funding for basic research that they justified as essential for society's defense and general well-being. Department of Defense (DOD) funding became ever more comprehensive and included a significant amount of theoretical work aimed at solving basic science problems. Projects were generously funded, with little direct oversight. Professors were able to work for the Pentagon without closely examining their relationship to the military.

However, the DOD did not fund in the name of pure science. Although many discrete projects may not have been concretely linked to weapon systems, the flow of funds to the university slowed down when the Pentagon felt its needs were not being met. In the mid-1950s, DOD considered jettisoning a host of "basic" projects to better attend to its operational needs. During the Vietnam War, the academic contribution to weapons systems was severely criticized, and funding was cut.[42] Only continued Cold War crises, such as those triggered by Sputnik and the "missile gap," kept research funds flowing. Though the university chose not to inspect closely the link between academic research and the service rendered by federal mission agencies, the two were often joined by war hysteria and the continued escalation of the arms race.

The Vietnam War forced the university to confront the relationships between research and defense. Students vehemently protested university research that furthered the war or filled the coffers of defense industries. The student movement won a number of victories. ROTC was moved off many campuses. Most universities banned classified research on the grounds that secrecy was inimical to the academic enterprise.[43] A number of universities ruled against faculty's engaging in clandestine work, such as Central Intelligence Agency (CIA) service, on the grounds that such activity, if discovered, would compromise legitimate scholars' research.[44] The 1970 Mansfield Amendment was a response to student protest. It forbade DOD support of research not directly and obviously connected to defense needs.[45] The Mansfield Amendment was aimed at weakening the tie between the military and the academy, and making each more accountable to the public.

Although the student movement of the 1960s forced the academy to examine its ties to the military, it also provided the consequences that university administrators and the scientific establishment had always feared. Scientists lost political power and influence; research funding declined precipitously. Nixon blamed his defeat in Vietnam at least in part on university-based opposition to the war. In what was widely interpreted by the academic community as a punitive gesture, he dismantled the presidential science advisory apparatus in 1973.[46] In the mid-1970s, DOD research monies for academic science fell from more than a quarter to less than 10 percent of the federal research and development budget.[47]

Instead of defense, funds for academic research in the 1970s went to energy, natural resources, environmental projects, and space research and technology.[48] However, this money, as the presidents of prestigious research universities never tired of pointing out, was "unpredictable," short term, too sensitive to the caprice of politics. University presidents sought more stable long-term funding commitments that would fulfill the same function defense research had—the provision of dependable resources that did not invite close oversight. Such a funding source was seen as essential to further basic research, on which university prestige rested.[49]

Change in the economy and the structure of science offered the university an alternative source of funding. In the early 1970s, the corporate community became enamored of high technology as a way to improve its position in the global market. At the same time, university science became a source of high technology. Discoveries in biotechnology, pharmaceuticals, computers, lasers, materials science, and robotics made basic research performed in the university more available for product development and marketing. The distinction between research and development, always somewhat artificial, seemed to be disappearing entirely.

Prominent university presidents, such as those who sit on the Business–Higher Education Forum, appeared ready to work with corporate leaders to put basic science to work for economic development. Indeed, high technology was presented to the public as a developmental imperative, essential to winning economic victories in the battle for global markets. University presidents such as Harvard's Derek Bok reminded the academic community that the social utility of knowledge has always been central to university missions and spoke of university-industry partnerships negotiated around the commercial exploitation of basic research as the "new" service.[50] In the words of the Business–Higher Education Forum, "The health of American basic research is critical in an era when international competition increases industry's need for scientific advances."[51] Research and development, basic and applied, academic and industrial, are the cornerstones of an economic development strategy aimed at continued U.S. dominance of global markets. Advanced technology, whether process or product, is viewed as commodity crucial to increasing declining corporate profitability and refurbishing depleted university resources.

The defense establishment too is convinced of the usefulness of research in high technology fields. This posture on the part of the DOD has had several consequences for the university. First, the DOD increased its emphasis on university research. In so doing, it was able to escape the constraints imposed by the student movement in the 1960s. The Mansfield Amendment was reinterpreted during the Carter administration to permit basic research to be funded by the Pentagon. And the DOD now takes its mission to be so broad that almost all scientific knowledge falls within its purview. In the words of the Department of the Army, its mission is "support [of] the national strategy in the face of any aspect of... threat."[52]

Second, the difficulties of accepting classified research are somewhat mitigated. Because a great deal of research is initially defined as basic, many DOD contracts are unclassified when they are let since their strategic importance is unclear. Instead, such judgements are made after the fact. Several agencies, led by the National Security Council, are attempting to instate procedures that allow research review panels to engage in de facto classification. Such a scheme reverses the classification process, permitting universities that currently prohibit classified research to accept contracts that might later be classified.[53]

Third, university involvement in military-related high technology has increased federal oversight of academic research. Sometimes the federal interest in research is expressed in outright surveillance and regulation. The Commerce Department's International Trade in Arms Regulations (ITAR) has been used to block the presentation of unclassified technical papers in areas such as photo-optical instrumentation because representatives of Soviet bloc countries were scheduled to be present. Similarly, Export Administration Regulations (EAR) have been used to prevent invited Eastern bloc representatives from attending U.S. academic meetings dealing with bubble memory.[54]

Basic academic research has become increasingly important to the development of strategies of corporate and military leaders. Each group is interested in global dominance in its own sphere. U.S.-based multinational corporations want greater shares of international markets. The defense establishment wants to hold the world balance of power. Each sees high technology as critical and academic research as a source for it. University presidents have been responsive to overtures from the corporate sector and the military, and have made propositions of their own to these groups, a subject that will be discussed in the next section. However, as universities become more involved in global development strategies of corporate and military establishments, research becomes increasingly subject to the restraints of geopolitics.

The issues surrounding the social goals of academic science are becoming more complicated as basic research is incorporated into high technology. The military continues to put forward ever more costly and complicated weapons systems, such as Star Wars. These are supported by some corporate leaders, especially those with large defense contracts, but others are beginning to articulate programs for civilian development of product-oriented high technology. As they do so, they become increasingly critical of military research and development programs, arguing that defense spending is capital intensive but nonproductive, a waste of scarce resources in a deficit economy. University presidents frequently call for increasing both defense and corporate spending, evading the debate emerging between corporate and military leaders. Professors seem divided on defense programs. Sometimes they oppose military funding, as is the case with the thousands of physicists who have taken a position against SDI. At other times, they participate in weapons development.[55] Generally, faculty seem quiescent about the growth of corporate interest in higher education.

Although not marked by protest and demonstration, as was the case during the Vietnam War, criticism of weapons development and the arms race probably comes from a much broader social base than in the 1960s. There is no strong student movement addressing these issues, but social movements, such as the Jackson campaign and the peace movement, are calling for the curtailment of defense programs, arguing against further development of nonproductive, capital-intensive, destabilizing, and destructive weapons systems. The forces represented by these movements also ask that less attention be paid to devising strategies for winning global markets and more to developing programs that aid domestic manufacturing and consumption.

There is presently a great deal of conflict around the setting of national goals for scientific research. Support for imperial goals remains strong among higher education policy makers, but whether these will be realized in the military sphere or global economic markets remains unclear. There is also considerable popular criticism of imperial ambition, whether arms or trade is involved. Interest surrounding the setting of science policy is unlikely to abate, given the amount of resources at issue.

Resources

The modern research university is resource dependent and looks to business and government for funding. Historically, the university's strategy has been to use one patron to increase leverage over the other, always increasing the resource commitment of both. Although absolute support on the part of both business and government has consistently increased since World War II, government bears by far the greatest share of university costs. Even if only academic science is considered, business falls far behind. In the early 1950s, business support accounted for 10 to 15 percent of the cost of academic science. By the early 1970s, it accounted for less than 2 percent. Business support has increased since the early 1970s, but even multimillion-dollar university-industry research contracts have not brought business's share of the costs much over 6 percent.[56]

Although corporate support is dwarfed by federal expenditures, the university community sees it as crucial. Increased corporate support lessens university dependence on the political process. When federal research monies failed to grow in the late 1960s and early 1970s, and ever more strings seemed to be attached to what monies there were, university presidents began to look for other sources of support, as well as ways of gaining greater leverage over government spending. An alliance with multination corporations engaged in high technology met both these needs. These corporations have significant resources to support research, and have the political skills and money necessary to lobby Washington effectively for greater government contributions to their own and university research programs. Indeed, the Business–Higher Education Forum itself is a concrete and practical example of an effort by managers of elite research universities to work with corporations to gain greater leverage on government resources.

In the last ten years, corporate leaders have committed millions of dollars to university research contracts in high technology fields. Among the more spectacular contracts are the Harvard-Monsanto Agreement (1974), a twelve-year, $23 million contract for basic cell research related to the growth of tumors; Exxon-MIT (1980), a ten-year, $8 million contract for basic research in combustion processes; DuPont-Harvard (1981), a five-year, $6 million contract for genetic research; Mallinckrodt–Washington University (1981), a three-year $3.9 million contract for basic research on "hybridomas," a technique for producing useful biological materials such as antibodies; and Celeanese-Yale (1982), a three-year $1.1 million contract for basic research on the composition and synthesis of naturally occurring enzymes.[57]

Corporate leaders are willing to commit some monies to research and development, but they use their policy networks to convince politicians that public investment in privately controlled high technology will bring the greatest possible return to society. In other words, their development strategy turns on "privatization," or socializing the costs of development, maintaining profits, and hoping that prosperity will expand to include the majority of the citizenry. The public is asked to spend increasing amounts of tax dollars to underwrite university-industry agreements.[58] Examples are the University of Delaware Center for Catalytic Technology, where one-third of the financing comes from industry, one-third from the National Science Foundation (NSF), and one-third from mission-oriented agencies; the Case Western Reserve Center for Applied Polymer Research, where NSF has supplied $750,000 for five years, and the university is making an effort to attract corporate sponsors to match these monies; and the University of Massachusetts–Industry Research Center on Polymers, where NSF granted $1 million over five years, attracting thirteen corporate sponsors, each contributing $20,000 a year to the Center.

In these industry-university-state partnerships, government funds are supposed to serve as seed monies. Private industry is supposed to match and eventually pick up the tab. Although it is probably too soon to evaluate these projects, thus far government expenditures account for more than the total industry contribution and a great deal more than any individual corporation. What happens at the federal level also occurs in state government. States like California and Massachusetts contribute monies to university-industry partnerships to retain their edge as high technology centers and maintain their economic base. States like New York and Pennsylvania contribute monies in the hopes of eventually developing a high technology economic base.[59] Again, public resources are used to underwrite private enterprise.

These expenditures are justified on the grounds that they rekindle the economy, finally benefiting the society as a whole. Although this theory of development has enjoyed wide support, there is mounting evidence that challenges it. High technology industries are capital intensive, not labor intensive, and are more likely to use small numbers of highly trained professionals than large numbers of blue-collar workers.[60] Even if high technology brings prosperity to a region, it frequently does not benefit the preponderance of the populational already there, which is often unskilled or skilled for heavy industry.

Privatization has a dimension beyond using public monies to increase private profit; it transfers what was once part of the public domain to the private. Scientific knowledge becomes intellectual property, and ideas cease to circulate freely. Increasingly, university-industry contracts are written to allow for patent application prior to publication.[61] According to a recent Supreme Court decision, genes developed through biotechnology are patentable, and therefore potentially profit bearing. On-line computer catalogues are introduced to research libraries, and for the first time so are fees for student searches.

University leaders are almost as interested in privatization and the creation of intellectual property as corporate leaders. As the cost of research soars and resources become more scarce, university presidents have begun to exploit the intellectual potential of their faculty through patents and contracts. Until the early 1970s, patents were generally held by universities only if they were nonexclusive, developed economically to meet broad public use. Now universities have exclusive patent policies that cover faculty discoveries, even though these are often made with the help of public subsidy.[62] Faculty are increasingly expected to market their own skills aggressively and to secure private as well as public contracts that are handled by the university grants and contracts office, with a share for the institution. University presidents seem to hope for a nonstop flow of royalties, licensing fees, and contract overhead to ease their financial problems.

However, direct federal commitments to high technology research have not been large. Instead, primary funding for corporate development and university research has come through DOD. Until the 1970s, from 50 to 80 percent of the national research and development budget went to DOD, and approximately half of that went directly to industry, mainly for development.[63] In the 1970s, the DOD share of the total federal research and development budget fell, briefly and dramatically, in response to widespread social protest. It dropped to 44 percent, with a corresponding loss of federal subsidy to industry.[64] Beginning in 1978, this trend was reversed. By 1985, defense was firmly in first place among mission agencies, far surpassing Health and Human Services. Industry shares are expected to rise until they account for as much as 70 percent of the DOD budget. Although the university will not benefit as greatly, its share too increases.[65]

Increased commitment to defense seems to be necessary to substantially raise public resources committed to both corporate and academic research and development. Cold War hysteria, arms race escalation, and the promise of defense-related jobs make these increases acceptable to the public. The defense work of corporations and universities overlaps in high-technology areas as they create the stuff of Star Wars: advanced computers, robotics, lasers, and materials sciences. Universities and corporations often share in defense-funded projects. For example, Defense Advanced Research Projects Agency (DARPA) has put $8 million into Standford's Center for Integrated Systems, which serves seventeen industrial sponsors, each contributing $250,000 per year; DARPA and the Office of Naval Research (ONR) are providing funds to supplement Westinghouse's $1 million, five-year investment in the Robotics Institute at Carnegie-Mellon.[66]

Corporate, military, and university interests have converged around high technology, which depends on applying basic science to industrial processes. Basic science is a resource all would command. It is presented to the public as the key to economic and military ascendancy. What is not explained clearly is the economic benefit to the ordinary taxpayer. Yet the public pays for high technology research not once, but two and sometimes three times over. It pays initially for the research and development—often whether in industry or the university—that makes high technology possible. The public pays again when it has to purchase the product for whose development it has paid. It pays indirectly through lost public revenue when corporations receive tax breaks for engaging in research and development. But the public receives no direct share of the very considerable profits corporations realize, and it is not certain whether the average citizen finally benefits indirectly from the "trickle down" effect.

The concentration of science resources on commercial and military high technology has several consequences for higher education. Institutional stratification is likely to be exacerbated. The research envisioned is so capital intensive that funds will be available only for a few institutions. Supercomputers and supercolliders illustrate this point nicely. The institutions likely to benefit from these projects are the ones that have gained historically from federal science monies: those designated as elite, AAU, Research I, or flagship institutions. Competition between these institutions will probably increase until state systems finally shake down to one or two research universities per state, whether private or public. These institutions will probably be characterized by greater specialization, which will be determined by how deeply a university is invested in a particular technology.

Because more monies will be concentrated on fewer elite research institutions, less may be available for expanding access and career opportunities. Instead, more and more students will be concentrated in low-cost community colleges. The numbers of first-generation college students, blacks, and Hispanics in two-year institutions will probably increase. Although the vast majority of community college students are in vocational-technical programs, "ladders" between two-year training and career preparation at elite research universities will become harder to build. The gap between the careers that generate high technology and those which use or service it will widen because the design and development of equipment, as well as the experience gained by working with it, will be too costly to replicate in the two-year system.

The effect of science-funding patterns on nonelite, four-year, independent sector institutions and state comprehensives is unclear. Perhaps these institutions will focus on preparing students for the quasi- or semiprofessions, such as accounting, business, marketing, personnel, and advertising in the private sector, and teaching, nursing, and social work in the public. Whatever direction the four-year schools take, given the concentration of science funds on fewer elite research institutions, they are unlikely to be able to modify their missions markedly or to expand greatly.

Concentration of resources on high technology will probably alter the shape of the traditional disciplines. Most high technology endeavors bring together scientists at the cutting edge of already specialized fields, as was the case with biotechnology. Traditional departments were very often unable to accommodate these working groups, since doing so meant more than incorporating a new specialization represented by one or two faculty members. As a result, extradepartmental units are being formed, which often report directly to the central administration. Very often these units erode the strength of traditional disciplines and weaken broad notions of colleagueship.

These new research units will probably not serve as sites for careers for women, blacks, and other minorities. High technology fields tend to draw from areas in science and engineering that are deeply rooted in sophisticated mathematics. For a variety of reasons, these areas have been difficult for groups historically excluded from higher education to enter.

Overall, greater institutional stratification with increased concentration of science resources means that fewer high-paying, high-prestige programs and career slots will be available. Since higher education is still the key to social mobility, that means mobility will be somewhat curtailed. As has been the case in the past, some groups will experience a great deal more constraint than others.

The Legislative Agenda of the Business-Higher Education Forum

For each of the policy issues discussed—access, careers, the social utility of knowledge, resources—there is a dynamic in which groups previously excluded make significant gains, which are then not fully realized. In the instance of access, working-class youth, blacks, and other minorities, were able to increase college attendance greatly, largely because of system expansion in the several states and sustained increases in student aid packages. However, the majority of ethnically and racially diverse students are increasingly concentrated in community colleges in programs not clearly connected to lucrative or prestigious careers. Despite gains in some areas, people of color and students from working-class origins are not frequently found in graduate school. Although more women are completing graduate school, they are concentrated in traditionally female fields, or frequently do "women's work" in male-dominated areas.[67] With regard to resources for national science agendas, monies that were put into environmental, renewable energy, and social or educational programs in the 1960s and early 1970s have been redirected toward the development of military or commercial high technology.

It is possible to attribute these changes to natural and inevitable swings in the political pendulum, to a popular conservative resurgence, as embodied by the Reagan administration, or a somewhat Marxian dialectic, in which popular forces struggle against the more fortunate. However, these explanations say very little about policy actors who have persistently taken higher education as a major issue, and even less about the institutions from which they draw their support. I think it is possible to gain the greatest understanding of higher education policy dynamics in the past two decades by seeing leaders of central social institutions—such as corporations and higher education—as systematically working to regain ground lost in the political upheavals and reforms that took place in the 1960s. I see these leaders as policy actors representing their institutions in a host of forums that cover a broad spectrum of issues. They preceded the Reagan administration in calling for the redirection of policy in ways that would create a climate better able to foster business interests and entrepreneurial science. For want of a better term, I will provisionally call these leaders "representatives of the institutional class," a nomenclature and concept that I will address more fully in the final chapter. I do not see the institutional class as consciously attempting to defeat or roll back gains made by previously excluded groups, but as attempting to regain some of the privileges and resources they have lost so that they can rechart a course to realize their vision of American prosperity. Although these groups do not seek directly to constrict educational opportunities for poor or working-class youth, women, blacks, or ethnic minorities, the legislative agendas they put forward often have that effect, especially in periods of scarce resources.

The location of corporate leaders in the institutional class is fairly obvious, but the inclusion of university presidents is less so. Certainly presidents of prestigious research universities often place different emphases on policy issues than corporate leaders and have a number of divergent interests; however, they are above all concerned with maintaining and enhancing the positions of their institutions. To do so, they have to form enduring political alliances around mutual concerns that gain them ready access to public monies while preserving their institutional autonomy. Joining with the corporate community through the promotion of entrepreneurial science is a major mechanism for creating such a strong political alliance.

The legislative agenda of the Business–Higher Education Forum points to policy issues promoted by the institutional class. The legislative agenda of the Forum can be treated in terms of four main categories: (1) tax cuts and revision of (2) antitrust and regulatory legislation as well as (3) intellectual property laws, and (4) the use of the state to organize business interests to provide resources for commercial products and processes. As I present the Forum's position on these issues, I comment briefly on likely but apparently unanticipated consequences.[68]

To begin with, the Business–Higher Education Forum asserts the primacy of the private sector in the single recommedation made in its initial report, *America's Competitive Challenge:* "As a nation, we must develop a consensus that industrial competitiveness is crucial to our social and economic well-being. Such a consensus will require a shift in public attitudes about national priorities, as well as changes in public perceptions about the nature of our economic malaise."[69] All decision making is subordinated to global economic success. Although economic development is important, this particular formulation of the problem calls for continued American economic dominance of world markets, in effect maintaining the hegemony achieved after victory in World War II. Alternative strategies of development, such as more limited growth, selective growth, or an emphasis on domestic manufacturing and consumption, go by the board. In a very real sense, national policy becomes business policy.

First, the Forum advocates a series of tax breaks that are designed to stimulate research by offering large deductions for research and development and greater write-offs for equipment donations to universities. These tax credits serve to bind industry and universities more closely together, since each depends on the other to realize gains in terms of projects, equipment, and write-offs. When these tax breaks are added to government expenditures for industrial, university, and military research and development, all of which directly or indirectly benefit corporations and the higher learning, the whole adds up to a massive subsidy for business and universities.

This pattern of indirectly funding research serves to exempt government and the university from the political process. The only electoral debate over research agendas supported by tax credits is on the appropriateness of the mechanism. Substantive questions about research agendas are not treated. Although this funding pattern escapes the electoral process, it is still subject to the bureaucratic process, with the definition of research and development to some extent falling into the hands of the Internal Revenue Service.

Second is the revision of antitrust and regulatory laws. Relaxing antitrust laws is seen as a way to economize on the cost of research. The Forum asks that corporations be allowed to combine for research and development. In effect, they pool resources, research, and results. Examples of such pooling agreements on which the Interstate Commerce Commission has already ruled favorably are the Semiconductor Research Corporation, the Microelectronics and Computer Corporation, and the University Steel Resources Center.

Although such requests sound reasonable, shared research would probably increase monopoly in product lines. Combination is unlikely to occur in highly competitive emerging industries with low-entry costs and unassigned patent rights where aspiring entrepreneurs could benefit. The pooling of resources and knowledge in these circumstances would create authentic competition, which monopoly sector corporations historically have sought to avoid. The pattern is for research consortia to emerge in monopolistic industries, where high cost and prior rights make possible the pooling of knowledge along with profits. For example, the petroleum industry, which regulates competition through oil leases and mineral rights, has long cooperated in research.[70] When young industries, such as computers, mature, they too begin to eliminate competition as costs rise and patents are proved, making attractive cooperative research at the enterprise level. The end result is more likely to raise the costs to the consumer than to lower them.

Antiregulatory legislation is advanced as a mechanism for freeing productive forces by doing away with excessive bureaucratic constraints. The Forum says that regulation must be relaxed, especially with regard to pure food, drugs, and environmental controls, if productivity is to increase. Although the Forum supports the expansion of equal opportunity for women and minorities, it nonetheless includes the state and federal affirmative action apparatus in its campaign to reduce regulation, arguing that self-regulation is less costly, less intrusive, and equally effective.

However, the costs and burdens that corporations allege they labor under are grounded more in rhetorical than in empirical observation. Very little empirical evidence sustains corporate cries of abuse.[71] What is really at stake is the voice that the public gained with regard to corporate personnel processes and corporate products through struggles in the 1960s and 1970s for affirmative action, occupational health and safety, and environmental protection. Victories were won with the National Environmental Protection Act of 1969, the Occupational Health and Safety Act of 1970, the Clean Air Act of 1972, the Water Pollution Control Act of 1972, and the Toxic Substances Control Act of 1976. Antiregulatory legislation is aimed at reversing these gains, bringing personnel and product decisions firmly under corporate control again, despite the social costs this might entail.

Third, the Forum advocates changes in intellectual property laws. In general, these changes are designed to increase private ownership. Patent law revision has made it possible for corporations and universities to patent and exploit privately discoveries made with public monies. Stricter control is also sought with regard to patents in the international market. Here, the object is to maintain private profit and a competitive edge with regard to new technological discoveries. Developing countries will have every opportunity to buy products, but technological processes will be more carefully guarded than ever before by export control and patent laws.

In terms of domestic development strategies, the Forum's intellectual property legislation supports greater use of the patent process. Patent historians almost uniformly point out that the primary use of patents has been to maintain exclusivity of manufacturing rights rather than to disseminate discovery broadly. Patent litigation undertaken by large corporations is notorious for impeding production for small firms with innovative modifications or alternatives to existing technologies.[72] In terms of international development strategies, the Forum's intellectual property legislation is likely to increase rather than mitigate disparities between first and third world countries. Although this strategy might benefit U.S. corporations in the short run, it may harm the development of countries we look to as potential consumers as we strive to increase our share in global markets.

Fourth, the Forum seeks to use the state to organize business interests for international competition and to fund these global endeavors. The Forum advocates the creation of a number of organizations at the federal level, among them an office for an advisor on economic competitiveness, with the same staff and status as offices that treat national security, science, domestic policy, and trade; a National Commission on Industrial Competitiveness; and a bureau to serve as an Information Center on International Competitiveness in the Department of Commerce. According to the Forum, "government's responsibility is not to direct the activities of the private sector, but to streamline its own processes and create an environment in which the individual and collective talents of the private sector can be focused to meet the competitive challenge."[73] Thus, the administrative arm of the state is the vehicle through which cooperation between sometimes contentious groups can be realized. "Business, education and labor must become more dynamic and flexible. In the past, a large number of specific recommendations have been made, each designed to solve a specific problem. The missing link has been an agreed-upon framework for the many actions, in the public and private sectors, that will constitute the American response."[74] The Forum legislative program is developed within a corporatist framework emphasizing sectoral representation by elites working outside the electoral process, and using the bureaucratic state as the vehicle for reaching consensus. Although labor is mentioned, the primary emphasis in Forum documents is on the partnership between the corporate and academic communities. The basis of industry-university cooperation is mutual agreement to seek state subsidies for pursuing entrepreneurial science. Although corporate and university leaders do not want the state to infringe on their autonomy, they nonetheless expect the bureaucratic state at the national level to help them realize their objective. In essence, business and higher education seek to use high technology to return to conditions in their respective institutional sectors similar to those in the 1950s, when each experienced untrammeled prosperity.

Conclusion

The higher education policy process in the postwar period can be accounted for with some plausibility by several theories—pluralism, neo-Marxism, and corporatism. Each interpretation has strengths and weaknesses; none accounts fully for the developments which have occurred. I will consider each briefly.

A pluralist interpretation of the postwar period would concentrate on the electoral process, political parties, and the proliferation of interest groups. Increased access would be explained in terms of the formation of political coalitions, usually intent on and able to influence Democratic party platforms, with confirmation of coalition positions at the polls. Expanded career opportunities would probably be seen as the logical corollary of greater access. The process of system expansion would be seen as initiated in the political process, and furthered by organized actors—the higher education associations, foundations, heads of pertinent congressional committees—in the higher education policy arena.[75]

Science policy and the shaping of research agendas, as well as the allocation of funds to mission agencies, would probably be treated separately from access and career preparation, and explained from a more nearly neopluralist perspective. Organized scientists speaking with the authority of expertise would be seen as using their knowledge as a resource to win access to the executive branch and to influence science policy. In this interpretation, the end goals of national science policy—national defense, energy self-sufficiency through nuclear power, the conquest of space—would not be questioned, and analysis would focus on shifts in influence between scientists, government, and leaders of the business corporations charged with development. The major policy questions would turn on how end goals are to be realized and how greatly each of the sectors represented by elites will benefit. The dynamics of policy would probably be explained in terms of positional authority and different levels of political skill and scientific expertise possessed by the various actors.[76] The Business–Higher Education Forum would be seen as one more variation in the combination and recombination of elite policy actors playing yet another round in the policy arena.

Pluralist theory is able to explain overall increases in access and career opportunities, but not systematic disparities in attainment on the part of groups such as working-class youth, women, blacks, and other minorities. It can also speak intelligently to the organizational mechanisms and political stratagems used to shape policy within the legislature and certain parts of the bureaucratic state. However, pluralist theory does not treat obvious disparities flowing from differences in class, power, and resources, and, as a result, does not look at the ways in which these patterned disparities influence policy formation. Moreover, it overlooks the ways in which issues are shaped and defined prior to their appearance in electoral politics and on state bureaucratic agendas.

A neo-Marxian interpretation of higher education policy formation in the postwar period would explain increased access in one of two ways. On the one hand, the power and omniscience of corporate leaders, acting as agents of a ruling class defined by its relationship to the means of production, could be emphasized, and higher education seen as a vehicle for cultural reproduction. In cultural reproductive terms, increased access would be seen as a mechanism for tracking students of particular class backgrounds into appropriate postsecondary sectors, reproducing both their class status and the division of labor.[77] On the other hand, changes in access and career opportunity could be taken as a manifestation of class conflict, in which groups distant from the means of production are able to organize and make some real gains, despite opposition from the upper class or its agents.[78]

Science policy would be seen as shaped primarily by corporate leaders who control the technological infrastructure. The military and university could be seen as part of the superstructure, responding relatively mechanically to the corporate sector. Alternatively, the military could be seen as a repressive state apparatus ready to enforce the will of the ruling class with violence, and the university as an ideological state apparatus engaged in constructing hegemonic social views that create definitions of reality that serve the dominant class. The degree of autonomy of the state from the dominant class, the mechanisms by which state and class structures are connected, and the permeability of the state apparatus by various class fractions would depend on the theorist. Although each theoretical permutation would have somewhat different consequences for interpretations of science policy, there would be overall agreement that the state mediates class interests and conflict.[79]

The Business–Higher Education Forum could be seen by neo-Marxists in any one of several ways. The Forum could be construed as business-as-usual, an instance of infrastructure determining superstructure, predictably shaping policy that favors the corporate class. The Forum could also be seen as an aggressive response by the dominant class to the popular politics of the 1960s and early 1970s, another turn in the dialectic. The Forum too could be taken as an effort on the part of policy actors representing the ruling class to insert themselves into bureaucratic state processes in a concrete instance of participation in mediation.

The neo-Marxian interpretation attends to power, especially that held by the dominant class, offers a way to account for disparities in achievement and the distribution of social rewards, includes more than electoral politics when treating policy formation, and begins to address the bureaucratic state as a site of enduring class and ideological conflict, not easily remedied by ameliorative reforms. Although notions of class conflict contribute to understanding policy dynamics, this theoretical tradition often gives too much weight to the dominant class, frequently attributes a progressivity and unity to the working class that does not seem to attend to the actuality of popular politics, and generally ignores the professional-managerial class, except when treating it as an appendage of the dominant or working class. As a result, the dialectic often plays out in unanticipated ways.

Corporatism is a way of explaining what happens when interest group politics result in political gridlock. At that point, established elites appeal to broader interests, grounded in production. Representatives are drawn from the major economic sectors, rather than being elected from geographical locations. The sectors vary but usually include business, labor, agriculture, and various professional groups. Leaders come together, usually on the initiative of business and sometimes under the aegis of government, to reach consensus on national policy issues. Agreement about policy is reached outside the electoral process. The policy produced is usually seen as bipartisan, and is expected to figure in electoral politics. The issue before the electorate is not the policy itself, but how much or how little support to allocate.[80]

In the United States, corporatism is basically a post-1960s theoretical development. As such, it does not address itself to issues of increased access or expansion of career opportunities. In terms of recent science policy, theorists would probably see the corporatist model as expanding the policy formation process to include more than the scientific, corporate, and government communities, as by-passing fragmented, contentious, and narrow interest groups, and as undercutting incipient class conflict by including labor. Although it does not use the term, the Business-Higher Education Forum takes a corporatist position on most issues, as examination of its legislative program indicated.[81]

Although corporatism can explain the enduring patterns of interest group and party politics in ways that pluralism cannot, and uses the notion of sectors in a way that comes closer to accounting for the relative strength of various classes in the policy process, it too has problems. In the United States, corporatism assumes, as does the Forum, that business interests should be paramount, and provides for no dynamic that would shift this balance of power. However well corporatism might address the moment, it is static and conservative. Although corporatism speaks to the state and its complexity, it sees the only viable role of the state as promoting business interests, such as global competitiveness. This accounts for neither the ways in which electoral politics interact with the bureaucratic state, nor for other state possibilities, such as the state as an autonomous sector, or the state as the site of class struggle. Finally, corporatism is inherently antidemocratic, which makes enduring legitimacy a serious problem.

CHAPTER 3

Social Location of Corporate and University Leaders

In this chapter, I look at the social origins, education, career paths, and activities of Business–Higher Education Forum leaders. This information should help us understand the routes by which these leaders came to hold positions in organizations concerned with policy formation. It can be used to test the explanatory power of the several theoretical traditions under consideration—pluralism, neo-Marxism, and corporatism—since the several theories all consider access to position and power. By looking at the Business–Higher Education Forum leaders in the context of these theories, we can perhaps begin to think of policy formation in broader terms than specific issues or particular branches of government. Taking a broader view might eventually help us see new and more consistent patterns of policy formation.

First, I will briefly review what the theories informing this work have to say about access to policy-making positions or power. Then I will describe the method and data developed in this chapter. Finally, I will appraise these theories in light of the data.

Pluralist theory maintains that social class and economic power are not the major determinants of political power. Instead, there are many routes to influence, although the primary mechanism is participation in the electoral process, via voting, political parties, and interest groups.[1] Neopluralist theory recognizes that some individuals have more power than others, and that leaders of large organizations or economic institutions have an especially strong voice in the policy process. But power is seen as rooted in formal position and offices rather than in class or wealth, and attention focuses on how positions of authority are won. Mobility is central. Neopluralist theory argues that willing and able individuals can attain these positions through outstanding educational and career performance. In the economic sector, for instance, theories of "managerial revolution" postulate that education and professional training are more important than ownership in securing top leadership positions in large corporations. Although the numbers of leaders who achieve positions with policy influence are relatively small, there is a broad possibility of access through a relatively open educational system.[2]

Marxist theory sees class, as determined by relationship to the means of production, as the key to power. The ruling class, or bourgeoisie, own the means of production, and are therefore able to dominate all political and social processes.[3] Neo-Marxian theory continues to argue for the importance of class position in determining access to power, but recognizes that ownership and management are no longer necessarily vested only in a family and its heirs, and concedes that ownership is diffused, with financial institutions often having the controlling interest in publicly held corporations. Neo-Marxists make a case for the continued centrality of class by noting the concentration of wealth in the hands of a relatively small number of families, or upper class, and contend that these families, through their multiple positions as stockholders, members of boards of trustees and foundations, and in some instances, as CEOs, corporate attorneys, or politicians, still exert a strong personal influence on policy formation. They are also thought to exercise power by virtue of their relationship to persons who serve corporations but who are not members of the upper class. This group is sometimes spoken of as a power elite, and can include CEOs, corporate attorneys, media persons, and various professionals. It is usually assumed to be linked directly or indirectly, but concretely, to the upper class, and is seen as a vehicle for realizing upper-class policy.[4]

Alternatively, the state can be seen as the vehicle which organizes the interest of the upper class. In this view, the systemic constraints of advanced capitalism are such that state and educational leaders are committed to enhancing corporate captial. If capitalism falters, the whole system suffers. The state is thought to mediate between various corporate groups, as well as between corporate and other groups, always representing the interest of the dominant class.[5]

Corporatism sees power as determined by sector. The major sectors are most often business, labor, agriculture, and the professions. Business has the greatest voice in determining policy, because a strong economy is critical to widespread prosperity. Although the most able are thought to rise to the top of any given sector, mobility between sectors is not viewed as a serious problem. Democratic considerations are attended to by including representatives of the several sectors in decision-making processes.[6] Thus, corporatism sees power as dependeent on position within the sector and does not attend closely to the chances of the individual.

Methodological Considerations

Power structure research was used in creating and analyzing this data set.[7] Biographical information, work histories, and social-class indicators were compiled from standard reference sources. The personal statistics collected were place of birth, age, marital status, and number of children. Educational information included the institutions that conferred the baccalaureate, the highest degree, and the field of specialization for the highest degree. Military service and career patterns were examined. The numbers of corporate boards on which leaders sat, as well as the ones they shared, were identified. The service leaders rendered was noted, whether they worked for the federal government, acted as trustees of universities and foundations, or were participants in policy institutes and business forum groups. Their efforts in professional and trade associations were catalogued, as was their membership in social clubs.

The sample of leaders was composed of all of the members of the Business–Higher Education Forum in 1983, the year the first national report, *America's Competitive Challenge*, was released.[8] The major source was *Who's Who in America*, supplemented by similar works that covered special fields and women.[9] Using *Who's Who* as a source is not without problems. Since the subjects of these brief biographical sketches fill in the standard forms themselves, there is no guarantee that information is complete, or even accurate. Pertinent data are sometimes not solicited. For example, father's occupation, probably the most reliable single indicator of social class, is not requested. Of the 81 leaders, biographical data were found on 77 (95 percent). Information was not obtainable on two corporate and two university leaders. The sample, then, consists of 39 CEOs and 38 university presidents.

The data were coded using Fortran, and cross-tabs were run. Throughout, the Forum leaders are discussed as a group and then CEOs are compared to university presidents. Finally, the data for each group are compared to studies done on similar groups.

Overall, the social-class background and leadership experience of corporate and university leaders were remarkably similar. There were exceptions—corporate leaders were older, had higher incomes, fewer advanced degrees, and belonged to more social clubs—but these were hardly unexpected. In general, both sets of leaders seemed to be the sons and daughters of the middle or upper middle class, attended college at rather undistinguished institutions, and graduate school at more exclusive ones. The CEOs usually made their careers in one or two companies, the presidents at three or more universities. Most shared responsibility for managing major social institutions other than their own, with corporate leaders serving as university trustees and university presidents sitting on corporate boards. Almost all were joiners, devoting a great deal of energy to furthering the interests of trade and professional associations, as well as policy, civic, educational, and economic groups. Their activities reflect their base in different sectors, with corporate leaders focusing on the host of institutions that create the infrastructure for the higher learning. However, what is most remarkable about their activity is the degree to which each set of leaders regularly participates in the routine business of the other.

Composition and Origins

The sample as a whole was composed largely of white males. Although all the CEOs were white men, the university officers included six women (15.7 percent of educational leaders) and one black. Four of the women presided over single-sex colleges. Gender and racial variation, then, were confined to the educational sector. Even there, women were usually represented in a separate sphere.[10]

The Forum leaders were well into middle age, with approximately 60 percent between fifty-three and sixty-three years of age. Corporate leaders were somewhat older than university presidents; 92.2 percent were born before 1930, as compared with 55.2 percent of university presidents. The oldest, a corporate leader, was sixty-seven, and the youngest, a university president, was forty-four.

The leaders tended to come from regions of the country outside the New England and Middle Atlantic States. Fifty-one (66.2 percent) were born in the South, Midwest, or Far West. Of these, 27 (69.2 percent) were corporate leaders and 24 (63.1 percent) were university presidents. Although both groups were more likely to have been born in urban areas than in rural areas, more corporate leaders (46.1 percent) were born in rural areas than were university presidents (28.9 percent). The region and population density of their place of birth probably reflect national demographic patterns of their age cohorts quite closely. However, the origin of many outside the metropolitan centers of the culturally and economically dominant New England and Middle Atlantic states suggests that the group as a whole was not the offspring of the upper class or established elites.[11]

The leaders tended to have very stable marriages. For those on whom marital information was available, 64 (83.1 percent) were married once, 7 (9 percent) were divorced and remarried, and 4 (5.1 percent) were single. Of corporate leaders, 33 (85 percent) were married once, and 5 (13 percent) were divorced and remarried. None were single. Of university presidents, 31 (82 percent) were married once, 2 (5.3 percent) were divorced and remarried, and 4 (10.5 percent) were never married. The higher rate of single persons among the university presidents is explained by the fact that three were Roman Catholic priests. The group as a whole produced an average of 2.8 children apiece. The corporate leaders averaged 3.4, and the university presidents 2.3. The leaders, then, seem to have been supported in their careers by traditional marriages.

Education

Almost all the leaders (93.5 percent) received a baccalaureate degree (see Table 3.1). As would be expected, more university presidents (97.4 percent) held B.A.'s than did corporate leaders (87.1 percent), but the differences between the two samples are not great, especially when the relatively small percentage of men graduating from college before 1950 is considered. Overall, 37 (48.0 percent) attended private colleges and universities, and 29 (37.6 percent) went to public institutions. University presidents were somewhat more likely to attend private schools than their corporate counterparts. Twenty (52.6 percent) of the presidents received their B.A.'s from private-sector institutions, as compared with 17 corporate leaders (43.5 percent). Given the fact that the corporate leaders were older, and most probably attended college before 1940, well before the expansion of public sector higher education, the relatively high percentage in state colleges and universities is somewhat unexpected.

The largest group of leaders attended the prestigious research universities that constitute Carnegie 1.1 institutions (28, or 36.3 percent).[12]

However, the same number (28, or 36.3 percent) attended other Class 1 schools or Class 2 institutions. Although other Class 1 institutions are also research universities, their status in the 1930s and 1940s was probably not the same as in 1976, when the Carnegie classification took place. For example, Rice, Auburn, and the Georgia Institute of Technology, all now classified as Carnegie 1.2, in the earlier period most likely lacked some of the indicators that would have resulted in a Class 1.2 rating.[13] And institutions which now fall in the Carnegie Class 2 category—North Carolina State University, Indiana State University—in the 1930s and 1940s were in many cases primarily concerned with teacher education. A fair number of leaders, then, attended relatively undistinguished Carnegie Class 1 and 2 undergraduate schools, many of which were public.

Only 7 leaders (9.0 percent) attended the selective liberal arts colleges in Class 3.1, which are by and large private. Although approximately half of the university leaders and roughly 40 percent of the corporate leaders were schooled at private colleges, relatively few attended the Ivy League or Seven Sisters. Only 11 (14.2 percent) received degrees from Ivies, with the largest number (5) concentrated at Harvard, and 3 had Seven Sisters baccalaureates. Only 13 (17 percent) attended those twelve institutions that control over half the resources available to private higher education.[14]

The prestige of the institutions from which leaders received their B.A.'s is quite mixed. Thirty-five (45.4 percent) held degrees from Carnegie Class 1.1 or 3.1 institutions. Even in these categories 10 (13 percent) of the degrees were from public institutions. Thirty-one (40.2 percent) held degrees from rather undistinguished institutions, many of which were public.

Thirty-eight (49.3 percent) of the 77 leaders received M.A.'s or M.B.A.'s. The CEOs held 15 (19.4 percent), and the presidents 23 (30 percent) (see Table 3.2). Of these degrees, 20 (55.5 percent) were won at private institutions and 16 (44.4 percent) at public. At this level, corporate leaders tended to select private institutions, and university leaders were almost evenly balanced between private and public. The institutions they attended were more likely to be in Carnegie Class 1.1 and 1.2 than was the case with the B.A., and by far the greatest number were in Class 1.1.

The master's degree serves many functions—as an initial professional degree, as a preliminary step to a Ph.D., or as a consolation prize for students judged not able to go forward with work on a higher degree.[15] The many purposes of the M.A. make it difficult to know what significance to attach to the numbers of degrees held and the institutions at which they were obtained. Perhaps what is most notable about M.A.'s in this sample is their increasing concentration at Carnegie Class 1.1 research universities.

Of the 77 leaders, 38 had earned Ph.D.'s (49.3 percent; see Table 3.3).[16] As would be expected, most were held by university presidents: 31, or 81.5 percent. The majority (27, or 71 percent) of the degrees were conferred by Carnegie 1.1 research universities. Most of these (19, or 70.3 percent) were privately controlled. Harvard was the institution where most (4) were concentrated. All but one, the California Institute of Technology, were among the twelve institutions that account for over half of the resources available to private higher education. All, including the eight public universities, were members of the Association of American Universities in the late 1940s, the time by which the majority of the sample would have completed graduate school.[17]

The next largest group of institutions granting Ph.D.'s to the leaders could not be classified in terms of public or private control, largely because the institutions were foreign. For the most part these were British—Oxford, the University of Birmingham, the University of Bradford. Another 6 (16 percent) leaders received Ph.D.'s across the remaining Carnegie Class 1 and 2 categories.

As the leaders moved upward on the educational ladder, CEOs dropped off, and university presidents were concentrated more heavily in elite, private research universities. Although university presidents tended to receive B.A.'s from private colleges, they were not necessarily elite schools, sometimes not even widely recognized institutions. As they took higher degrees, however, they were more likley to receive them from institutions at the apex of the research and resource establishment.

Professional degrees present a somewhat different case. Ten leaders had professional degrees. Seven had passed the bar, 4 corporate and 3 university leaders. but none indicated they had attended law school. Apparently they "read" for the bar instead, an option no longer open to most students. The 3 medical degrees, all held by university presidents were conventional, from elite, private institutions.

The field of specialization of the highest degree varied quite markedly by group (see Table 3.4). Fourteen (35.8 percent) of the corporate leaders and 8 (21 percent) of the presidents did not report a field of specialization. When they reported, they were concentrated in quite different fields. CEOs tended to be engineers or hold business degrees, and university leaders held degrees in the social and physical sciences, with a strong concentration in the social. The only social science field that corporate and university leaders shared was economics, the dismal science. Both corporate and university leaders were scattered across a wide range of physical sciences, sharing but a single field, geology.

More than half (42, or 54.5 percent) of the leaders served in the military, with the navy as the preferred branch (18, or 23.4 percent; see Table 3.5). The CEOs, who were more of an age for World War II, signed on more frequently than the presidents. Twenty-seven CEOs, or 69.2 percent, were in the military, 23 (58.9 percent) during World War II. The highest concentration was the wartime navy, where 14 (35.8 percent) served, most as officers. Only 4 corporate leaders were in the armed services after the war. Fifteen, or 39.5 percent of the university presidents were in the military. Most served after World War II, with the heaviest concentration in the army. The military was clearly a central experience for the corporate leaders, perhaps the most common institutional bond among them. For the university leaders, the military seems not to have provided a collective background but did provide shared training for a fair number.

Making any sort of definitive statement about the social origins of this sample of corporate and university leaders is difficult, since central pieces of data—father's occupation and income—are missing. However, it is possible to make some inferences from data on origins and education. The roughly 40 percent of the sample born outside of large cities, when contexted with the roughly 70 percent born away from the economically and culturally dominant New England and Middle Atlantic regions, suggests that a good number of the leaders probably came from the small towns and regional cities of America. Twenty-seven (35 percent) were of an age to attend college before World War II. Their families were most likely located in the comfortable classes if they felt able to forego their sons' labor or pay fees for college, even at relatively inexpensive state schools. In 1940, only 18 percent of the eighteen-to-twenty-one-year-old group attended college.[18] However, the choice of undergraduate colleges, with slightly over half attending institutions other than Carnegie Class 1.1 or 3.1 institutions, suggests that these parents were not overly or uniformly prosperous. Those who were able to send their sons and daughters to prestigious private schools were probably located in the upper middle class.

The increasing concentration of leaders at elite institutions as they made their way up the educational ladder may be due in part to the GI Bill. If it is assumed that World War II disrupted the leaders' educational experience, most would not have attended graduate school until after 1945. Since roughly half (42, or 54.5 percent) served in the military and 92.2 percent were male, GI benefits were readily available. Graduate students could have relied on the government rather than their families of birth, again suggesting that elite social origins were not crucial to their success.

The literature of social mobility and status attainment does not contradict the interpretation I offer on social origins. The social mobility literature indicates that intergenerational mobility occurs within a very narrow range of occupations, and in small steps. Very few persons are born to the families of farmers or unskilled workers and rise to the presidency of a large corporation or university.[19] Or, as Thomas Dye says when discussing his sample of institutional elites, "Even those who have experienced considerable upward mobility are likely to have risen from middle- or upper-middle-class families rather than working class families."[20]

The status attainment literature stresses the importance of education in career choice. According to the Wisconsin model, parents' socioeconomic status and individual ability are the starting point in status attainment, but education is a significant variable, having a strong independent effect on the early phases of a person's career.[21] The heavy investment of leaders in higher education may begin to serve as an explanation of their initial movement from middle-status families to positions of power and authority.

In sum, the leaders seemed to be drawn mainly from the small towns and regional cities of America, and were probably from comfortable middle- or upper-middle-class families. There is little indication that any were from the working class, or, conversely, that many were from upper-class families or established elites. What is unusual about the group is their military service and their access to educational benefits following the war, which may have provided this leadership cohort with more mobility than most.

Careers

The majority of corporate and university leaders had careers confined to a single sector. Thirty-one CEOs (76.9 percent) made their careers exclusively in the business world; 28 (71 percent) of the presidents spent their professional lives in the university (see Table 3.6).

There was not a great deal of variation in the career pattern of CEOs who worked only in corporate settings. Twenty-nine moved from professional or managerial status within their firm to CEO. Two were workers, who moved from the factory floor to CEO. This career pattern is not unexpected, confirming the literature that suggests there is little possibility of movement between nonmanagerial and managerial tracks within firms.[22]

The careers of presidents who worked only in the university were almost as predictable. Twenty-five moved from research associate or professor to managerial status, usually as dean, and then on to a presidency, chancellorship, or head of an educational agency. Two repeated this sequence, to the point of a deanship, more than once; one started as an assistant dean, became a professor, and then a president. Overall, a professorship in a well-established discipline still seems to be the starting point for the presidency of a major university.

Six corporate leaders had experience outside the business world before becoming CEOs. Most (5) were university professors who became executives. One moved from the military to a corporate presidency.

Eight university presidents had experience outside colleges and universities. One was a CEO who became a university president. Six moved between government and institutions that service the postsecondary sector before becoming presidents. Their government jobs were usually at the federal level, and they most often worked as high-ranking civil servants. Two more followed a similar pattern, but also worked for philanthropic foundations before becoming university presidents.

Three leaders, 2 CEOs and 2 university presidents, had careers so mixed they were impossible to classify. They worked across all sectors and also held political positions.

Corporate leaders demonstrated remarkable stability in the course of their careers. The majority (51.2 percent) made their entire career in a single corporation (see Table 3.7). Three-quarters worked at no more than two corporations. University leaders were much more mobile in the course of their careers. Most (60.5 percent) worked in three or more institutions on their way to the presidency. Leaders who confined their career to a single university were often at the most prestigious institutions.

The overwhelming majority of corporate and university leaders made careers in the world of either business or education. When leaders moved from one sector to the other, professors were more likely to move into the business world than were corporate leaders to the university. However, corporate leaders certainly had experience with higher learning, since almost all held B.A.'s, and many attended graduate school. Indeed, CEOs have probably had more experience in the realm of education than have university presidents in the world of commerce. However, university leaders were more likely than corporate to have experience with government or foundations.

Interlocking Directorships

The 77 leaders sat on the boards of 107 corporations (see Table 3.8). Of those who were on boards, most were on two or three. As would be expected, more corporate than university leaders sat on boards. Twenty-eight CEOs (71.7 percent) were on boards as compared with 21 university presidents (55.2 percent). A greater number of presidents confined their activity to a single board than did CEOs, who were more likely to be on two or more.

If the corporations on whose boards leaders sat are classified according to Fortune 500 categories, most served on industrials. The corporate leaders were on 38 (57.5 percent) of these, and university leaders on 21 (51.2 percent). The next largest category for CEOs was banks, with corporate leaders sitting on 11 (16.6 percent) of these, as compared with only 4 (9.7 percent) for university leaders. However, university leaders were more likely to sit on the boards of financial corporations (6, or 14.6 percent) than CEOs, who sat on only 2 (3 percent). The rest of the corporations on which the leaders sat were scattered throughout the remaining Fortune 500 categories, with no notable concentration in any one.

Of the 107 corporations on which the 77 leaders sat, there were direct connections between 43 (40.2 percent), excluding the 9 boards on which the corporate leaders served as member as well as CEO. Twenty-one university presidents sat on 41 boards, of which 8 (19.5 percent) were the boards of CEOs in the Business–Higher Education Forum. Twenty-eight corporate leaders sat on 66 boards, of which 6 (9 percent) were each other's. CEOs shared 4 boards outside of each other's, with 3 serving on a single board, Bank America Corporation. CEOs and presidents shared 5 boards, with 3 serving on a single board, Chase Manhattan Bank. University presidents shared 4 boards, with 3 sitting on a single board, Ford Motor Company.

Although fewer university presidents than CEOs sat on corporate boards, over half (55.2 percent) sat on at least one. Approximately 20 percent of the university leaders served on corporation boards headed by Business–Higher Education Forum members. Overall, there were direct connections via board membership among 40.2 percent of the leaders. Although the Forum membership is hardly defined by alliances formed in corporate board rooms, there are enough attachments to argue for shared opportunities for social intercourse that might facilitate policy formation in settings outside of corporations.

Institutional Leadership

As leaders are members of corporate boards, so they are trustees of other major social institutions. Leaders regularly sit on the boards of colleges and universities, foundations, and policy institutes. Their service is to some degree predicated on the sector to which their careers are bound, but in many cases is interchangeable.

Thirty-three leaders (42.8 percent) were trustees, overseers, directors, regents, or visiting board members of postsecondary institutions. Most (23, or 29.9 percent) were trustees of a single institution, and 20 (25.9 percent) were trustees of two or more. Corporate leaders (28, or 71.7 percent) were more frequently trustees than university leaders (15, or 39.4 percent). Fifteen CEOs (38 percent) were trustees of a single institution; 13 (33.3 percent) were trustees of two or more. Fifteen university presidents (39.4 percent) were trustees of other institutions of higher learning, with 7 (18.4 percent) serving on more than one. Corporate leaders were very often trustees of university presidents' institutions. Sixteen, 41 percent of all CEOs and 57 percent of those who were trustees, presided over Forum presidents' universities.

As university presidents sit on boards of corporations, so corporate leaders serve as trustees for colleges and universtities. The rather high percentage of corporate leaders on universtiy presidents' boards suggests that the halls of higher learning, like corporate conference rooms, are places where Forum leaders establish personal ties while guiding central social institutions. Although each group is active in the other's sphere of interest, the corporate leaders tend to be somewhat more involved than university presidents as the formal custodians of both business and education.

Twenty leaders were directors of major philanthropic foundations. Among the foundations were the Carnegie Foundation for the Advancement of Teaching, the Twentieth Century Fund, the Rockefeller foundation, and the Danforth. There was no noticeable concentration of leaders on the board of any one foundation. University presidents were more heavily involved than corporate leaders. Fourteen presidents (36.8 percent) served as trustees of foundations, as compared with 6 CEOs (15.3 percent).

Leaders also served as trustees or directors of university foundations, organizations concerned with higher education, or the boards of preparatory schools, art museums, hospitals, and policy institutes. I will briefly examine the distribution of leaders in those groups of organizations in which ten or more leaders were active.

Fourteen leaders (18.1 percent) served as trustees of organizations that constitute the infrastructure of the postsecondary sector. These organizations are neither colleges nor universities but are essential to the functioning of educational institutions. Among the organizations were the College Entrance Examination board, the College Retirement Equities Fund, the Council on Post-Secondary Accreditation, and the United Negro College Fund. Six CEOs (15.3 percent) and 8 university presidents (21 percent) were trustees for this category of institution. Corporate involvement seems high, given that the skill, experience, and expertise needed to work with institutions serving the postsecondary sector are more likely to be possessed by educators than business people.

Sixteen leaders (20.7 percent) sat on hospital boards. Most were CEOs (11, or 28.2 percent). However, 5 university presidents (13.1 percent) also served on such boards.

Thirty leaders (38.9 percent) held 57 memberships on policy institutes or forums (see Table 3.9). Seventeen corporate leaders (43.5 percent) served in 34 policy organizations, and 13 university presidents (34.2 percent) were in 23. There tended to be a division of labor with regard to policy, with businessmen concentrating on organizations concerned with business or economic policy and university presidents working with foreign policy groups. When there was overlap, university leaders were more likley to participate in business or economic groups than CEOs were to work with foreign policy organizations.

The groups with the largest concentration of leaders, ten or more, were the Business Roundtable, the Committee for Economic Development, and the Council on Foreign Relations. The Business Roundtable is a relatively new organization, created in 1973 by the merger of two anti-labor groups. The Roundtable "has adopted the aggressive role of pressing for national legislation through direct participation of its nearly 200 members—each of whom is a chief executive officer of one of the nation's largest corporations.[23] It became a formidable lobbying organization for the business community, actually using CEOs to present Roundtable positions forcefully on Capitol Hill. In general, the group is "anti-regulation, anti-federal spending, anti-anti-trust, pro-tax cuts."[24] The Committee for Economic Development is an older organization, founded in the early 1940s to shape policy in the postwar era. The corporate leaders who founded the Committee saw themselves as shaping plans for the future that would reflect the national rather than special interests, but in a fashion that was acceptable to the business community. The Committee was composed of 200 corporate leaders, who were later joined by a small number of university presidents.[25] The Council on Foreign Relations was founded in 1920-1921 by bankers, lawyers, and professors who realized that the United States would play a greater role in world affairs after World War I. It is the oldest of the policy organizations and has the largest membership— 1,500. About half of its members are from the East Coast, and half from the rest of the nation. Large banks and corporations are heavily represented, but there are also professors and university presidents.[26]

All, of course, are members of the Business–Higher Education Forum. As an organization, the Business–Higher Education Forum is more akin to the Committee on Economic Development or the Roundtable than to the Council on Foreign Relations. The Forum is small, and is composed exclusively of persons in the top leadership positions of their organization. Like the Roundtable and the Committee, the Forum is concerned more with economic issues rooted in domestic institutions than with the political ramifications of foreign policy. As was the case with the Roundtable in the business sector, the Forum was organized in the 1970s to cope with a political climate perceived by leaders of established insitutions to be more responsive to new claims advanced in the name of the public interest than to traditional concerns of the business and postsecondary community. Although substantially more corporate than university leaders serve on the Roundtable and the Committee, one-third of the leaders are presidents of universities. These organizations were probably a model for the Forum, and probably places where corporate and university leaders previously shared the work of shaping policy.

Leaders also served the federal government, with university presidents reporting a much higher level of activity than corporate. Three CEOs (7.6 percent) indicated that they worked with the government, with most service being rendered to the Department of Defense. Eighteen university presidents (47.3 percent) were engaged in service, holding 34 positions, 21 with presidential commissions or ad hoc bodies, and 14 with executive branch agencies. Education, science, and social problems dominated the concerns of the commissions and ad hoc bodies, and the Department of Defense, followed by the State Department, were the agencies in which the presidents served most frequently.

Although all leaders were active in an array of organizations outside their primary institution, corporate leaders seemed even more active than university leaders, except with regard to foundations and government service. The obvious reasons for including corporate leaders on boards is their ready access to resources, both economic and political, as well as their management skills and knowledge of the economy. In the case of philanthropic foundations, monetary resources are not a central concern, and in this area university leaders, who are familiar with managing the academics who frequently hold grants and have a need for philanthropic monies, are more highly represented than corporate. In the case of government service, university leaders may have a greater willingness to serve than corporate leaders and a greater need to demonstrate their expertise, especially in terms of higher education.

I also looked at leaders' participation in organizations with a more regional cast: higher education, business development, business booster, cultural and civic, music and youth groups. Only two groups of organizations had ten or more leaders involved: business development groups and business booster organizations.[27] Regional business development groups were primarily focused on high technology: for example, the Georgia Science and Technology Commission, the Illinois Governor's Commission on Science and Technology, and the California Economic Development Corporation. Twelve leaders (15.5 percent) were involved, 7 corporate (17.9 percent) and 5 university (13.1 percent). The rather high percentage of presidents probably is a function of the high technology emphasis of these regional development councils, with a concomitant emphasis on science. Business booster groups are local chambers of commerce, local business councils, urban leagues, and the like. Nineteen leaders (24.6 percent) were members of local booster organizations, 12 corporate (30.7 percent) and 7 higher education (9 percent). As in the case with business development organizations, corporate leaders dominate in booster organizations, but university presidents constitute a significant presence.

Overall, the leaders were involved in the leadership of a wide spectrum of social institutions. Corporate leaders tended to participate more frequently than university presidents and to concentrate their activities in the business sector. Nonetheless, they were very active in educational affairs, particularly as university trustees. And the presidents too were active in many business sector organizations. Perhaps what is most notable about their stewardship is the routine involvement of each group in the other's sphere of interest.

Trade Associations, Professional Associations, and Scholarships

In addition to contributing to the worlds of business and education, leaders were active in organizations that pertained to their special interests, be these commercial or professional. Corporate leaders tended to participate in trade associations, such as the American Iron and Steel Institute, the U.S. Industrial Telephone Association, and the Motor Vehicle Manufacturers Association. But they also were members of a number of organizations, mostly engineering, that represented their professional and commercial interest simultaneously: the Society for Automotive Engineers, the American Institute of Chemical Engineers, and the National Society of Professional Engineers. University presidents tended to continue membership in their disciplinary or professional associations: the American Economic Association, the American Statistical Association, the American Board of Medical Specialties, and the American Bar Association. But they were also active in a wide variety of organizations that served the postsecondary sector. And there were also a number of organizations in which both corporate and higher education leaders were members.

As a group, the majority of the leaders particiapted in professional and trade associations: 67 (87.0 percent) were members (see Table 3.10). University leaders joined more frequently than corporate, with 36 (94.7 percent) reporting membership as opposed to 31 CEOs (79.4 percent). Presidents tended to belong to two or more organizations, and corporate leaders were more likely to belong to one. The average number of associations in which university leaders participated was 3.3; for corporate leaders it was 1.6.

Corporate leaders did not overlap greatly in terms of their memberships, although there were some organizational clusters. Three CEOs belonged to the American Iron and Steel Institute, 3 to the Institute of Electrical and Electronics Engineers, and 3 to the Society of Automotive Engineers. Although there were more clusters of university leaders, they were not in the associations of learned disciplines. Only two of these had more than 2 university leaders as members: the American Economic Association, with 4, and the American Political Science Association, with 3. Instead, university presidents shared organizations concerned with the management of the postsecondary sector. Four reported membership in the American Association for Higher Education, 4 in the Association of American Universities, 3 in the American Association of University Professors, 3 in the American Educational Research Association, 3 in the National Association of Independent Colleges and Universities, and 3 in the Association of American Rhodes Scholars.[28]

Although the two groups of leaders were usually members of organizations that reflected their career-based interests, they did share membership in some organizations, most of which serve as quasi-public oversight agencies for their educational and science concerns. Thus, 15 leaders, 4 corporate and 11 university, were members of the American Council of Education; 9 leaders, 3 corporate and 6 university, were members of the American Association for the Advancement of Science; and 8 leaders, 1 corporate and 7 university, were members of the American Academy of Arts and Sciences. Although university leaders have a much heavier representation in these organizations than corporate leaders, both groups have a presence.

University leaders' ongoing concern with their disciplines is revealed in other ways than by their association membership. Their publication record also demonstrates a dedication to scholarship. Twenty-four presidents (63.1 percent) wrote 86 books. Nine authored 1 or 2, eleven authored 3 to 5, and four authored more than 5. In contrast, two corporate leaders report writing books, one a single volume; the other, 4 volumes. As the university presidents are clearly differentiated from corporate leaders in terms of Ph.D.'s, so they show an ongoing connection with their disciplines and an enduring commitment to scholarship. Although the worlds of business and education leaders are close on a number of dimensions, there are nonetheless clear areas of difference.

Social Clubs

Corporate leaders were much more heavily involved in social clubs than university presidents (see Table 3.10.) Approximately 75 percent of the CEOs belonged to social clubs, whereas only about 30 percent of the presidents did.[29] When the presidents were members, they tended to belong to one or two clubs, whereas the CEOs were more likely to be members in three to six clubs. There were a number of clubs in which 3 or more leaders were clustered, most of whom were corporate leaders. The Detroit Athletic, Pacific Union, and Racquet Clubs each had 3 leaders, all of whom were CEOs but one. The Bohemian Grove, Chicago, Fox Chapel, and Laurel Valley country clubs each had 4 leaders, all of whom were CEOs. Five leaders, all CEOs, were members of the Duquesne Club.

According to scholars who deal with social class, membership in the majority of these clubs (Bohemian Grove, Chicago, Duquesne) is an indication of upper-class status. The remaining two clubs—Fox Chapel and Laurel Valley—are, like the Duquesne Club, in Pennsylvania, and may indicate the presence of a close-knit group of corporate leaders moving in the same social circles.

Leaders' Social and Career Context

I will now compare these leaders to leaders in other studies that treat business and university leaders, to see if the Business-Higher Education Forum leaders are representative of their several communities. In each case, I am drawing on work that uses either large or summative data sets. I am contexting the university presidents in the work of Cohen and March, and the corporate leaders in that of Dye, and Useem.[30]

Cohen and March did not review the social-class background of the presidents in their sample of 42, but did thoroughly review the literature on presidents to create a context in which to discuss leadership. They relied primarily on four studies, as well as a cross-sectional, historical review of the men and women who held the presidency at the institutions they studied.[31] I will concentrate on their most recent data, which were collected between 1965 and 1970,[32] and which still remain the best on the "presidential careers" of university leaders.[33]

Cohen and March begin by noting that "the American college presidents today and in the recent past are most commonly middle-aged, married, white, Protestant academics from a relatively well-educated, middle class, professional-managerial, native born, small-town family background."[34] When the same characteristics are examined, this description as aptly fits Business-Higher Education Forum presidents as those in Cohen and March's study. The average age of the presidents in Cohen and March's study was fifty-three; it is approximately the same for the Forum presidents. As was the case with Forum leaders, almost all were married, and "virtually all unmarried presidents are members of celibate orders."[35] There were more women presidents in the Cohen and March study, 8 (19 percent), all of whom were heads of women's colleges, as compared with 6 (7.7 percent) in the Forum study, 4 of whom were presidents of women's colleges. When looking at other than historically black institutions, there were only 2 obviously black presidents discussed in Cohen and March's work, 1 of whom appears in the Forum sample.

According to Cohen and March, between 75 and 80 percent of the presidents now have Ph.D.'s; 81.5 percent of the Forum sample had doctorates. Humanities, education, and religion were important sources for presidents historically, but the social sciences were increasingly represented in Cohen and March's work, particularly at larger schools. The social sciences were the greatest source of Forum presidents' degrees (36.8.7 percent), followed by science (15.8 percent). The humanities, education, and religion were all represented, but each field accounted for no more than 5.2 percent of their degrees.

The work experience reported for presidents by Cohen and March was almost exclusively academic; according to one study, 41 percent had no experience outside the university, and 96 percent had less than five years. In the Forum sample, 77 percent worked only in the academy. Their single-minded focus on higher learning "does not distinguish them from most other administrators. Business executives, military commanders or medical chiefs of staff are also without significant experience outside their respective fields."[36] As is the case in Cohen and March, the vast majority of the Forum presidents began their climb toward the highest office from a professorship.

The presidents in the Forum sample look very much like presidents in general. There are some exceptions. The Forum presidents include fewer women, are more heavily drawn from the social sciences and the sciences, and have careers more tightly confined to the university than those reported by Cohen and March. The presidents in the Forum sample may well be from a somewhat more distinguished social-class background than the majority of presidents. As Cohen and March note, variations from the general population appear to be "related systematically to variations among colleges and universities in their student clientele and faculty personnel."[37] Since the institutions which the Forum presidents represent tend to be prestigious, as are their student bodies and faculty, with roughly 60 percent heading schools that are members of the Association of American Universities and 40 percent in Carnegie Class 1.1, it is likely that their backgrounds are not incongruent with the clientele they serve.

Unfortunately, Cohen and March did not provide information about presidents service on corporate boards or about the amount of time they spend participating in forums that try to shape national policy for education. There was no category of activity that treated presidential membership in groups like the Association of American Universities, the American Association for the Advancement of Science, or the Business–Higher Education Forum, or activities such as testifying before Congress or working with ad hoc commissions on higher education. Nor do they deal with the government service of presidents. Thus, the seemingly extensive work by Forum presidents in these areas cannot be compared with that of presidents generally.

Cohen and March argue that the prepresidential careers of university leaders provide a powerful socialization. They see presidents as absorbing academic values as students, as professors, and as administrators over a course of twenty to thirty-five years before becoming presidents. They see presidents as committed to academic values, but note that the socialization of presidents is different from that of faculty. The commitment of the Forum sample of presidents to academic values will be considered in the next chapter, and will be compared with both the values of business leaders and faculty in the final chapter.

Thomas Dye created a data base of 7,314 institutional leaders across corporate, public interest, and government sectors. I will draw on his subsample of those who hold the CEOs position in the top 100 industrial corporations, since this most closely matches the Forum sample of corporate leaders. The average age of Dye's sample was sixty-one; of the Forum, fifty-nine. The percentage of women in Dye's was 2.4 percent; there were no women in the Forum sample. Three percent of Dye's sample had no college, as compared with 10.2 percent of the Forum CEOs on whom there was no information on college attendance. Forty-three percent of Dye's sample had college degrees, whereas 89.8 percent of the Forum leaders held B.A.s. Law degrees were won by 22.6 percent of the top hundred industrial CEOs, and by only 10.2 percent of the Forum sample. However, 56.4 percent of the Forum sample held advanced degrees, as compared with 29.8 percent of Dye's.[38]

In terms of the type of school attended, 25.3 percent of Dye's sample went to public schools, compared with 41.0 percent of the Forum.[39] Eighteen percent of Dye's sample went to private schools, whereas 43.5 percent of Forum leaders attended them. Dye created a category of presitgious colleges and universities, composed of the twelve institutions that account for over half of the wealth controlled by private schools. Of the leaders who attended private schools in his sample, 54.9 percent went to these. In the Forum sample, 41.1 percent of the CEOs attending private schools received degrees from these institutions.

Most of Dye's leader's were urban born, as were 53.9 percent of the Forum sample. Dye estimated that 10 percent of the CEOs in his sample inherited their position, and that 30 percent of his entire sample was from upper-class origins.[40] Approximately half of Dye's top elites belonged to one or more of the list of thirty-six social clubs developed by Domhoff; 53.8 percent of the Forum sample belonged to one or more of the same clubs. Although club memberships are good indicators of current social status, they do not necessarily reveal much about social origins. Given the lack of information about the Forum leaders' family of birth and secondary schooling, it is impossible to compare them with Dye's sample in terms of class origin.

Dye does not develop data on the number of corporate boards on which the CEOs of the top 100 industrials sat, but he does present data on interlocks within groups. He uses Allen's data on 65 corporations in the 10 principal interlock groups in the United States to examine the argument for polyarchy, based on geographical proximity, with interlocked groups being dominated by large commercial banks.[41] Ten corporations (25.6 percent) in the Forum sample are in the interlocked groups, with the highest concentration in groups 3 and 4, dominated by the Mellon National Bank and the Morgan Guaranty Bank, respectively. The 10 corporations maintain interlocks with 197 others. In addition, Forum CEOs sit on the boards of 6 other corporations in the 10 principal interlock groups, maintaining through these corporations another 117 interlocks. Four of these are banks central to the 10 interlock groups. Forum leaders are dominant in only one of the 10 groups, represented on three of five boards dominated by the Mellon National Bank and centered in Pittsburgh.

As Dye did not develop data on the number of boards on which his top 100 industrial CEOs sat, neither did he provide systematic information on their government service. However, he notes that 15 percent of his entire sample are interlockers holding two or more top positions, with multiple interlocks—persons holding six or more positions—seemingly holding influential positions in policy networks. The majority of Forum CEOs are interlocked with other corporate leaders, university trustees, and policy groups, but are not close to government in any formal sense.

In 1977, Michael Useem developed a sample of 212 of the largest firms in the United States, which had a total of 3,105 senior executives and directors. Fourteen (35.8 percent) of the Forum firms were in Useem's sample.[42] Useem argues for an "inner circle," a "distinct, politicized business segment, if a segment is defined as a subset of class members sharing a specific location with partially distinct interest," that is concerned with shaping a policy climate in which general business interests will flourish. For Useem, the inner circle is defined primarily by those who serve on the boards of several corporations. In his sample of 212 firms, 267 sat on two boards, and 267 on three or more. In other words, approximately one-fifth, or 18 percent, of the entire set of managers and directors in his sample were inner circle members. Forum CEOs were more likely to sit on two or more boards. Eleven sat on two, and another 11 sat on three or more, for a total of 22 (56.4 percent), or somewhat over half of the leaders.

Indeed, in almost all the categories where comparable data were available, Forum leaders seemed to exhibit more inner circle characteristics than Useem's. Of Useem's managers, 10.9 percent of those who sat on the board of a single corporation were members of one or more of 15 clubs indicating upper-class social status, and 16.6 percent of the Forum members sitting on a single board were members. When Useem's managers sat on two boards, 31.5 percent were members of these clubs, and when they were on three or more boards, 46.9 percent were members. The percentages for similarly positioned Forum members are 54.5 percent and 63.6 percent respectively.

In terms of business policy-making organizations, neither Useem's managers nor Forum managers who sat on a single corporate board were very active. But those who sat on two or more were quite active, with Forum CEOs usually being more active, with one exception. Of the managers in Useem's sample who sat on two boards, 5.6 percent were members of the Business Roundtable, 7.1 percent of the Business Council, 9.0 percent of the Council for Economic Development, 12.7 percent of the Conference Board, and 15 percent of the Council on Foreign Relations. The percentages for Forum members who sat on two boards are 27.2, 27.2, 27.2, 45.4, and 0, respectively. Of Useem's managers who sat on three or more boards, 15 percent were members of the Business Roundtable, 19 percent of the Business Council, 12.8 percent of the Council on Economic Development, 23 percent of the Conference Board, and 15 percent of the Council on Foreign Relations. Twenty-seven percent of similarly situated Forum members were members of the Roundtable, 9 percent of the Business Council, 27.2 percent of the Council for Economic Development, 27.2 percent of the Conference Board, and none were on the Council for Foreign Relations.

All told, the 38 Forum leaders held 26 positions on these business-oriented policy formation groups. The only organization in which they were consistently less active than the managers in Useem's sample was the Council on Foreign Relations. As noted earlier, university leaders were much more likely than corporate presidents to occupy themselves with foreign policy concerns, so Forum members are quite involved in the Council, although not the CEOs.

Among CEOs who held membership in the Business Roundtable as well as two or more directorships, Useem found that 50 percent served on a federal government advisory committee. The leaders in the Forum sample, although active in the Roundtable and other policy groups, and often holders of two or more directorships, do not seem to engage in service at the federal level.[43]

In terms of holding trusteeships of universities, Forum leaders were significantly more likely than Useem's managers to serve on boards of postsecondary institutions. Of those managers in Useem's sample holding one corporate board seat, 5.6 percent were university trustees. Seventeen percent of those on two corporate boards were trustees, and 20.8 percent of those on three or more. The comparable percentages for the Forum CEOs are 66.6, 45.4, and 81.8. Generally, the Forum leaders are very highly involved in higher learning, much more so than comparable managers. The university on which the greatest number of Forum CEOs served was Carnegie-Mellon, with five.

Overall, Forum CEOs seem to be well integrated into interlocked corporations. The majority are probably positioned in the same way as leaders of most companies, where the head of the "typical firm is no more than three steps in the interlocking directorate from three-fifths to nine-tenths of all other substantial corporations."[44] There is some evidence that a group of Forum CEOs are dominant in a Pittsburgh-based, Mellon Bank–dominated interlocked group, evidence which is strengthened by the rather high number of forum leaders who serve on the Carnegie-Mellon University board of trustees, an insititution which is also closely related to the Mellons and is Pittsburgh based. In terms of the number of directorships held and a variety of other memberships, the Forum sample seems centered in the inner circle that Useem describes as a broad group of corporate leaders "who share with other corporate managers a common commitment to enhancing company profits [but whose] heightened sensitivity to business interests more general than those that look solely to support individual company profits also sets them apart."[45]

But connectivity in and of itself says little about the role CEOs play in policy formation. The data developed in this chapter indicate that

Forum CEOs are in a position to work jointly to influence national policy, but do not indicate their impact. However, it is possible to say that Forum leaders are well positioned to influence policy.

There are many differences between CEOs and university presidents in the Business–Higher Education Forum in terms of social origins, career, and public interests. CEOs, as a group, are older. They majored in fields such as engineering and business much more frequently than university presidents and did not take advanced degrees at anything approaching the same rate. They usually shared World War II and military service. They make much more money than the presidents; salaries for CEOs in the top 100 industrial corporations average about $500,000 before bonuses and stock options are calculated.[46] The corporate leaders are more often directors of a greater number of corporations than university leaders and are more involved in the business community and business policy organizations. They also belong to a greater number of exclusive social clubs. However, they do not engage in service to the federal government at the same rate as university presidents. The university leaders are differentiated from the corporate leaders by their possession of the Ph.D.; their involvement in foundations, learned societies, and higher education associations, and their commitment to scholarship. They were active in the military, but not at the same time or to the same degree or in the same branches as the corporate leaders. They seem to be much more interested in foreign policy and much more active in working for the federal government than CEOs. Generally, each group is differentiated from the other by commitment to careers in separate sectors.

However, there are many similarities among the differences. And the similarities may disclose the points at which the interest of the two groups converge. Each group heads central social institutions, and each serves as steward for a number of others. The corporate leaders display a rather unusual commitment to higher education. When compared with other CEOs, the Forum leaders are much more likely to serve as university trustees. They also are concerned with organizations that maintain the infrastructure of postsecondary education and with science policy groups. Higher education may be their avocation. Although the university presidents do not demonstrate as great an interest in business as the corporate leaders do in higher education, they show a consistent concern. They sit on corporate boards and maintain a presence in business organizations and business policy groups. Indeed, the two groups routinely participate in each other's sphere of interest, and probably regularly meet each other in a series of board rooms: corporate, university, foundation, business booster, business policy, postsecondary, and broad-based science policy forums. In the course of a year, a fair number of these leaders may encounter each other several times, usually when sharing the task of governing institutions other than their own, or when making policy in organizations that attempt to influence the political and economic climate in which the institutions operate.

The point of juncture between the two groups may be their interest in higher education, policy, and the creation of a climate in which the institutions for which they are responsible can flourish. The CEOs are concerned with higher education; the university presidents are active in policy organizations and government service. The Business-Higher Education Fourm may be the outcome of their shared interests.

Conclusion

When the data presented in this chapter are considered in light of the several theoretical traditions—pluralism, neo-Marxism, corporatism—no one tradition seems to explain the data perfectly. Instead, each seems able to account for some aspects of the data and not others. Let us briefly review the data from the perspective of the theories.

Pluralism does not seem to contribute greatly to an understanding of the data presented in this chapter. The forum is not representative of a broad social spectrum. Its members do not concentrate on electoral politics, but direct their political energies toward participation in elite policy formation organizations. The associations in which the leaders participate are not open to everyone; they are highly selective.

Neopluralism provides a somewhat better explanation. Education was clearly critical to leaders' careers, and the leaders' high attainment could be taken as a testament to their ability as well as systemic opportunity for mobility. However, the educational processes in which they were involved were generally not broad enough to include representatives from working-class origins, persons of color, or women in any considerable numbers, even though public institutions played a substantial role in the leaders' rise to positions of authority. Nor would neopluralism explain well the participation of the same leaders in such a wide spectrum of policy organizations. At most neopluralism allows for concentration of interests in the economic sphere. Other areas— foreign policy, cultural, educational—should show a greater independence.

In terms of Marxist theory, there was little evidence that any significant proportion of Forum leaders were members of an upper class. Nor was there evidence that a disproportionate number of CEOs owned the companies they headed. Because a large number of Forum leaders came from rural areas, where members of an industrial upper class are unlikely to be concentrated, and initially attended relatively undistinguished colleges and universities, which are unlikely places for an upper class to congregate, they are probably not contiguous with an upper or ruling class which derives its power from a relationship to the means of production.

The social origins of Forum members, which are probably disproportionately in the middle or professional classes, could accommodate that stream of neo-Marxian thought which stresses power and managerial elites as agents of the upper class, especially if fairly direct connections between Forum members and an upper class could be made. The data presented leave room for the possibility of such connections in some instances. A significant proportion of forum CEOs belong to social clubs thought to be the habitats of the upper class. However, it has not been convincingly demonstrated that persons with social club and upper-class membership are able or interested in performing as a ruling class. Forum leaders also belong heavily to organizations—corporate and policy—in which members of an upper class active in policy making are expected to participate. However, upper-class dominance of these organizations has not been conclusively shown. Although there is a possibility that Forum members may serve as agents or the ruling arm of an upper class, these data are not sufficient to address the question fully, much less speak to the question of the articulation with and the instrumentality of an upper class.

The neo-Marxist structuralist approach is less concerned with the origins of individual policy actors and more concerned with how the logic of the institutions they serve articulates with the overarching dialectic of capitalism. However, the structuralist position would seem to call for predictable connections between Forum leaders and the state, which neither group seems to display to great degree. Moreover, the state should be organizing the class interests of the Forum members, rather than Forum leaders trying to influence the state in their capacity as external actors.[47]

Nor does the social location, careers, and policy activity of Forum leaders seem to conform closely to a corporatist model. The Forum does not include broad sectoral representation. The most notable omissions are labor and agriculture. Although labor is occasionally mentioned as a key sector in Forum documents, there seem to be very few members of working-class origins. Nor are organizations that represent labor included in the broad spectrum of activities in which Forum members participate. Of course, there is a possibility that sectoral representation with labor and agriculture is not recorded in biographical materials, and occurs elsewhere.

Although each of the theoretical traditions can partially explain this body of information, none speaks directly to it. The data show leaders of two sets of central social institutions who appear to be connected closely to each other's institutions and to other institutions important to policy formation in general and to higher education in particular. These sets of leaders seem to be drawn from fairly similar social origins, which include neither the upper or lower reaches of society. Nor is there much evidence of sustained and direct ties to a ruling class or to the state.

This chapter has looked at the social origins of leaders, their education, career paths, and activities. These data allowed us to look at the paths by which leaders reached positions of authority in corporations and universities, as well as at the many other organizations to which they belong. Their work in the Forum can be seen as part of their broader participation in policy formation networks that speak to the institutional sectors they represent. However, these data speak only to the positions they hold, and say nothing about the power they wield or the substantive positions they take on various policy issues. In the next chapter, university presidents' fortunes in the policy process will be examined, as well as the substantive positions they advance with regard to the higher learning.

Table 3.1. Leaders' B.A. by Carnegie Classification[1] and Control

Carnegie Classification	Private Corporate	Private University	Public Corporate	Public University	Other Corporate	Other University	Total
Carnegie 1.1	9 (23.0)*	9 (23.6)	5 (12.8)	5 (13.1)			28 (36.3)
Carnegie 1.2	4 (10.2)	0	5 (12.8)	1 (2.6)			10 (13.0)
Carnegie 1.3	1 (2.5)	2 (5.2)	3 (7.6)	1 (2.6)			7 (9.0)
Carnegie 2.1	1 (2.5)	2 (5.2)	2 (5.1)	3 (7.8)			8 (10.4)
Carnegie 2.2	0	1 (2.6)	0	2 (5.2)			3 (4.0) 7

Carnegie 3.1	2 (5.1)	5 (13.1)			(9.0)
Carnegie 3.2	0	1 (2.6)			1 (1.2)
Carnegie 5.9**		1 (2.5)	1 (2.6)		2 (2.5)
Unclassified				2 (5.2)	4 (5.2)
Foreign			2 (5.1)	2 (5.2)	2 (2.5)
No Information on B.A.			0	1	5
			4 (10.2)	(2.6)	(6.5)
	17 (43.3)	20 (52.3)	16 (40.8)	13 (33.9)	6 (15.3)	5 (13.0)	77 (99.6)

Note: Percentages shown in parentheses.

[1] Carnegie Council, *A Classification of Institutions of Higher Education* (Washington, D.C.: The Carnegie Foundation for the Advancement of Teaching, 1976, rev. ed.).

*Percentages are of all corporate leaders (N = 39) and all university leaders (N = 38) respectively. The bottom line percentages are the percentages of all corporate leaders who received BAs from private sector postsecondary institutions, and the same for the number of university leaders, etc. across the several categories. If every other column in the bottom-line but the last is added up, they total 99.4% for corporate leaders, and 100.1% for university leaders. The last column is the percentage of all the leaders (N = 77).

**U.S. Service Academies.

Table 3.2. Leaders' M.A. by Carnegie Classification[1] and Control

Carnegie Classification	Private Corporate	Private University	Public Corporate	Public University	Other Corporate	Other University	Total
Carnegie 1.1	9 (23.0)*	6 (15.7)	3 (7.6)	5 (13.1)			23 (29.8)
Carnegie 1.2	0	1 (2.6)	1 (2.5)	3 (7.8)			5 (6.4)
Carnegie 1.3	1 (2.5)	1 (2.6)	1 (2.5)	1 (2.6)			4 (5.1)
Carnegie 2.1	0	1 (2.6)	0	2 (5.1)			3 (3.8)
Carnegie 3.1	0	1 (2.6)					1 (1.2)
Unclassified					0	1 (2.6)	1 (1.2)
Foreign					0	1 (2.6)	1 (1.2)
No Information on M.A.					24 (61.5)	15 (39.4)	39 (50.6)
	10 (25.5)	10 (23.5)	5 (12.6)	11 (28.6)	24 (61.5)	17 (44.6)	77 (99.3)

Note: Percentages shown in parentheses.

[1]Carnegie Council, *A Classification of Institutions of Higher Education* (Washington, D.C.: The Carnegie Foundation for the Advancement of Teaching, 1976, rev. ed.).

*Percentages are of all corporate leaders (N = 39) and all university leaders (N = 38) respectively. The bottom line percentages are the percentages of all corporate leaders who received MAs from private sector postsecondary institutions, and the same for the number of university leaders, etc. across the several categories. If every other column in the bottom line but the last is added up, they total 99.6% for corporate leaders, and 96.7% for university leaders. The last column is the percentage of all the leaders (N = 77).

Table 3.3. Leaders' Ph.D. by Carnegie Classification[1] and Control (Private or Public)

Carnegie Classification	Private Corporate	Private University	Public Corporate	Public University	Other Corporate	Other University	Total
Carnegie 1.1	3 (7.7)*	16 (42.1)	1 (2.6)	7 (18.4)			27 (35.0)
Carnegie 1.2		1 (2.6)	1 (2.6)	1 (2.6)			3 (3.8)
Carnegie 1.3				1 (2.6)			3 (3.8)
Carnegie 2.1				2 (5.1)			2 (2.5)
Unclassified					2 (5.2)	3 (7.8)	5 (6.4)
No Ph.D.					32 (82.0)	7 (18.4)	39 (50.6)
	3 (7.7)	17 (44.7)	2 (5.2)	11 (28.7)	34 (87.2)	10 (26.2)	77 (99.5)

Note: Percentages shown in parentheses.

[1] Carnegie Council. A *Classification of Institutions of Higher Education* (Washington, D.C.: The Carnegie Foundation for the Advancement of Teaching, 1976, rev. ed.).

*Percentages are of all corporate leaders (N = 39) and all university leaders (N = 38) respectively. The bottom line percentages are the percentages of all corporate leaders who received PhDs from private sector postsecondary institutions, and the same for the number of university leaders, etc. across the several categories. If every other column in the bottom line but the last is added up, they total 100.1% for corporate leaders, and 99.6% for university leaders. The last column is the percentage of all the leaders (N = 77).

Table 3.4. Leaders' Highest Degree by Field of Specialization

	N/A	ENGINEER	BUS	SOC SCI	SCI	HUM	LAW	EDUC	RELIG	MD
CEO	14	12	6	3	2	0	2	0	0	0
UNI	8	0	0	14	6	2	2	2	2	2
	22	12	6	17	8	2	4	2	2	2

Table 3.5. Leaders' Military Service

	N/A	WW2 ARMY	WW2 NAVY	WW2 AIR	WW2 MARINES	ARMY	NAVY	AIR FORCE	MARINES
CEO	12	5	14	3	1	0	3	0	1
UNI	23	1	4	1	0	6	1	2	0
	35	6	18	4	1	6	4	2	1

Table 3.6. Leaders' Career Patterns

CORPORATE CAREER		
Corporate only		
Manager to CEO		29
Factory floor to CEO		2
Other		
Military to CEO		1
Professor to CEO		5
Mixed		2
	TOTAL:	39
UNIVERSITY CAREER		
University only		
Professor to president		25
Professor to president, repeat		2
Manager to president		1
Other		
CEO to president		1
Government, postsecondary, president		6
Postsecondary, foundation, government president		1
Mixed		2
	TOTAL:	38

Table 3.7. Number of Institutions in Leaders' Career Path

Number of Institutions	Corporate leaders	University leaders
1	20 (51.2)	8 (21.0)
2	10 (25.6)	7 (18.4)
3	2 (5.1)	10 (26.3)
4	4 (5.1)	7 (18.4)
5	1 (2.5)	4 (10.5)
7	1 (2.5)	2 (5.2)
8	1 (2.5)	0
	39 (99.6)	38 (99.8)

Note: percentages shown in parentheses.

Table 3.8. Leaders' Representation on Corporate Boards

Number of Boards	Corporate leaders	University leaders	All
0	11 (28.2)	17 (44.7)	28 (36.3)
1	6 (15.4)	11 (28.9)	17 (22.0)
2	11 (28.2)	4 (10.5)	15 (19.4)
3	7 (17.9)	4 (10.5)	11 (14.2)
4	3 (7.6)	1 (2.6)	4 (5.1)
5	1 (2.5)	0	1 (1.2)
6	0	1 (2.6)	1 (1.2)
	39	38	77

Note: Percentages shown in parentheses.

Table 3.9. Leaders' Representation on Policy Institutes or Forums

Organization	Corporate leaders	University leaders	All
American Council on Capital Formation	1	0	1
Brookings	0	1	1
Business Council	5	1	6
Business Roundtable	7	3	10
Chamber of Commerce	3	0	3
Council for Economic Development	7	4	11
Conference Board	6	1	7
Council on Foreign Relations	1	9	10
Foreign Policy Association	1	0	1
Foreign Diplomats	1	0	1
Institute for Strategic Studies	0	2	2
Council for U.S.-China Trade	1	1	2
Overseas Development Council	0	1	1
Trilateral Commission	1	0	1
	34	23	57

Table 3.10. Leaders' Trade and Professional Associations

Number of associations	Corporate leaders	University leaders	All
0	8 (20.5)	2 (5.2)	10 (12.9)
1	14 (35.8)	2 (5.2)	16 (20.7)
2	8 (20.5)	13 (34.2)	21 (27.2)
3	5 (12.8)	8 (21.0)	13 (16.8)
4	2 (5.1)	3 (7.8)	5 (6.4)
5	2 (5.1)	5 (13.1)	7 (9.0)
6	0	2 (5.2)	2 (2.5)
7	0	1 (2.6)	1 (1.2)
8	0	1 (2.6)	1 (1.2)
9	0	1 (2.6)	1 (1.2)
	39 (99.8)	38 (99.5)	77 (99.1)

Note: Percentages shown in parentheses.

Table 3.11. Leaders' Representation in Social Clubs

number of clubs	Corporate leaders	University leaders	All
0	10 (25.6)	26 (68.4)	36 (47.7)
1	2 (5.1)	6 (15.7)	8 (10.3)
2	3 (7.6)	2 (5.2)	5 (6.4)
3	5 (12.8)	2 (5.2)	7 (9.0)
4	6 (15.3)	2 (5.2)	8 (10.3)
5	6 (15.3)	0	6 (7.7)
6	4 (10.2)	0	4 (5.1)
8	2 (5.1)	0	2 (2.5)
12	1 (2.5)	0	1 (1.2)
	39 (99.8)	38 (99.5)	77 (99.1)

Note: Percentages shown in parentheses.

CHAPTER 4

University Presidents and Public Policy

Ease of access to power is a point of disagreement among the several theories under consideration. Pluralists see individuals as able to join political parties and interest groups and influence policy, if not as individual actors in their own right, then through the leaders, candidates and officials who head the organizations in which they participate. Neopluralists recognize that influence may be more elusive, that the policy playing field is not level, and that some organizations, especially powerful and resource rich organizations, are able to tilt the field to their advantage. However, neopluralists do not see powerful organizations or interest groups as completely dominating the policy process and generally believe that readjusting the boundaries of the playing field will relieve inequities of power.[1]

Neo-Marxists see policy as shaped primarily by the dominant class, although they allow some room for resistance on the part of class fractions on which the dominant class would work its will. The professional middle class is generally thought to derive its strength from alignment with the dominant class, although some theorists think the professional middle class has a degree of autonomy and, under certain circumstances, can act in its own interests in the policy process.[2] Generally, the part of the working class remains undefined in any systematic way in the policy process, since this class is far from the seat of power and influence.

In corporatist theory, access to policy is gained by rising to leadership in organized socioeconomic producer groups that reflect the social division of labor. Presumably, the criterion for leadership is determined sectorally, with CEOs rising to the top on the basis of balance sheets in the black, and university presidents on the strength of the prestige of their institutions, which in turn rests on research, scholarship, grants and contracts. In labor organizations, leadership would probably stem from possession of skills useful for negotiating with leaders from other sectors. Through cooperative mutual interaction, sometimes under the aegis of the bureaucratic state, sometimes in cooperation with state bureaucrats, leaders of these peak associations are able to reach consensus on policy issues that reach across their several domains. In a sense, the power of corporatist structures depends on the ability of groups often in contention with each other to reach an accord.[3] However, corporate leaders in key enterprises probably have a stronger voice in determining policy than leaders from other groups because leaders in all sectors are thought to view economic productivity as central to maintaining social solidarity.

Examination of policy positions taken by university presidents over an extended period of time should allow us to see which of these theories best accounts for the substantive positions they take, the political alliances they make, and the degree to which their positions reflect those of other groups. At issue here is the degree to which presidents act as an independent interest group, as subservient to business groups, or as partners with leaders from sectors other than their own. Another question that will shed light on the several theories is the posture they take toward the state: do they see themselves as relatively independent external actors, pressuring the state; as external actors who form alliances with other organizations to pressure the state; as having their interests so intertwined with those of the dominant class that they are difficult to separate; as holding leverage on the state because they preside over institutions which serve as ideological and technical state apparatuses; or as able to shape state policy because of their ability to achieve broad policy consensus on specific issues among contending groups?

Methodological Considerations

Because all the theories expect organized action by institutional leaders in the policy process, I looked for the group that best represented the university presidents. To look at presidents' positions over time, I needed to locate a group that preceded the Forum, which has been active only for a short period. Moreover, I wanted a group separate from, but connected to, the Forum so the independent evolution of policy positions by presidents could later be compared to those developed jointly with corporate leaders in the Forum. Only two higher education associations met that criteria: the American Council on Education, and the Association of American Universities. Although the American Council on Education is the organization to which most university presidents in the Forum belong, I decided against it because, as a higher education umbrella organization, it represents all sectors of postsecondary education rather than those institutions most heavily involved in the Forum. Instead, I chose the Association of American Universities (AAU), which represents elite research universities, the institutional sector to which the largest number of presidents in the Forum belong. Approximately 56 percent of Forum presidents are AAU members. Although this is not a compelling majority, it is significantly greater than that of any other postsecondary sector, the next largest of which is selective liberal arts colleges, accounting for 20 percent. No other postsecondary sector accounts for more than 10 percent of Forum higher education membership. The AAU, then, has a history separate from the Forum, represents the sectoral interests of the largest number of presidents in the Forum, and is connected to the Forum through its membership.

The AAU is probably the most prestigious and exclusive of all higher education associations. It was "founded in 1900 by the fourteen American universities that then offered the Ph.D. degree." Currently, it has 54 American and two Canadian universities as members, all "with strong programs of graduate and professional education and scholarly research."[4] At present, half of the Association's members are public institutions, and half are private. An institution is represented by its chief executive officer, and membership is by invitation only. New institutions are considered for membership every three years, and their acceptance requires the assent of three-fourths of the AAU membership. The universities in the AAU are at the top of the postsecondary status hierarchy in terms of almost every indicator used in assessing quality and prestige: number of Nobel and other prizes and awards held by faculty, number of articles published in world journals, number of patents filed, amount of research dollars awarded annually by all sources, quality rankings such as those conducted by the American Council of Education, test scores of entering students, career achievement of graduates, size of endowments, and size of library holdings.[5]

I encountered a difficulty when looking for data sets which contained policy positions taken by the AAU: it does not keep a public record of its meetings, nor does it regularly publish a bulletin or journal. Interviews were not a viable alternative, given that there were a large number of presidents, widely scattered, many of whom were no longer in office. Therefore, I decided to create a data set of policy positions taken by AAU presidents by looking at their testimony before Congress over a fifteen-year period. The presidents were easily identifiable, and their testimony was indexed in the public record by name. Thus, I was able to create a uniform data set which included the policy actors in whom I was interested, and which addressed itself to questions of public policy in higher education at the national level over a fifteen year time span.

Although this data set represents presidents of AAU institutions, there is some question as to whether or not it represents the AAU. I cannot address this issue with the certainty I would like. However, the presidents' testimony does seem to give some evidence of reflecting the organized interest of research institutions. In 60 instances (35 percent) of their testimony, the presidents spoke as official representatives of higher education associations (see Table 4.1). Far and away the largest group represented a coalition of research universities. Many of these presidents opened their testimony with the following words:

> "I am testifying on behalf of the American Council on Education, the Association of American Universities, and the National Association of State Universities and Land Grant Colleges. The member institutions of these associations constitute the recipients of virtually all of the federal funds which are provided for academic science."[6]

Table 4.1. Associations Officially Represented by AAU Presidents in Their Congressional Testimony

Association Represented	Instances of Presidential Testimony*
Association of American Universities, National Association of State Universities and Land Grant Colleges, American Council of Graduate Schools or some combination thereof	39
AAU, NASUALGC and American Association of State Universities	5
National Association of Independent Colleges and Universities	5
Other Combinations of Above	3
Coalition Groups	3
Other	5
	60

*Total N. = 173

In 39 instances, or 65 percent of those in which presidents identified themselves as representing higher education associations, the presidents saw themselves as speaking on behalf of this coalition. This coalition represents most institutions with graduate schools, and is probably dominant in shaping policy in the higher education associations, considering its prestige, exclusivity, and activity in the other associations.

AAU presidents were identified through the *Educational Directory of Colleges and Universities* for the years 1970–1985. During this time span, 140 men and one woman served as presidents of the AAU universities. Their names were looked up in the *Congressional Information Service,* an index of testimony. The presidential testimony before the various Congressional committees was analyzed in terms of: frequency of testimony; committees to which testimony was provided; whom the presidents represented in their testimony; issues on which they testified; and the positions presidents advocated. The data were coded in terms of topics, themes and emphasis. Topics were coded in terms of the total subject a president addressed. Thus, there was only one topic per instance of testimony. Themes refer to issues and coherent approaches presidents brought to their topics. The coding unit for themes was paragraphs. There was no limit to the number of themes coded per testimony, and thus it was possible to capture some of the variety, argumentation and concerns of the presidents. Emphasis refers to the quality of language employed in discussion, allowing attention to tone and nuance. Emphasis was not systematically coded, but employed as an analytical tool in looking at the coded material. The topics, themes and emphasis are the basis for the discussion that follows.

In reporting on the topics and themes in the body of the chapter, I did not present the testimony of each president who addressed a topic or spoke to a theme. Instead, I selected those instances of testimony which I deemed to best represent the topic or theme under discussion. The frequency of testimony and number of presidents involved for any particular topic or theme can best be grasped by consulting the tables in the text.

Of the 141 presidents of AAU universities between 1970–1985, 64 (45 percent) testified before Congress. Of the 64, 20 spoke only once (see Table 4.2). The remaining 44 were quite active, testifying from two to eight times, and on average 3.5 times. In total, there were 173 instances of testimony.[7]

By far the greatest share of testimony was given before the House (see Table 4.3). AAU presidents spoke there in 116 (67 percent) in-

Table 4.2. Instances of Presidential Testimony

President & Institution	70	71	72	73	74	75	76	77	78	79	80	81	82	83	84	85	Total
Appley, Mortimer Clark University												1					1
Beering, Steven Purdue University															1		1
Bernstein, Marver Brandeis University							1										1
Bok, Derek Harvard University							1		1				2	1		1	6
Bowen, William Princeton University									1				2		1		4
Boyd, Willard University of Iowa					1	1					1						3
Brademas, John New York University													2		1	2	5
Brown, Harold Calif. Instit. of Technology			1				1										2

Byron, William Catholic University of America						1	2	2	5
Corbally, John University of Illinois		2							2
Corson, Dale Cornell University & Emeritus	1	1					1		3
Cyert, Richard Carnegie-Mellon University			1			1	1		3
Danforth, William Washington University				1			1		2
Eggers, Melvin Syracuse University		1	2	2	1	2			8
Enarson, Harold Ohio State University						1	2	2	5
Ferguson, Glenn Clark University	2								2
Fleming, Robben University of Michigan		1							1
Friday, William University of North Carolina	1					1	1		3

(continued)

Table 4.2. *continued*

President & Institution	70	71	72	73	74	75	76	77	78	79	80	81	82	83	84	85	Total
Giametti, H. Bartlett Yale University																1	2
Goheen, Robert Princeton University			1							1							1
Goldberger, Marvin Calif. Instit. of Technology												1	1				2
Gray, Hanna University of Chicago													2				2
Gray, Paul Mass. Instit. of Technology														1	1	1	3
Hackney, Sheldon Tulane University/Penn								1						1	1	1	3
Hanson, Arthur Purdue University												1					1
Harrington, Fred University of Wisconsin		1															1

Heard, Alexander Vanderbilt University					1					1
Hereford, Frank University of Virginia										1
Hester, James New York University	1	1		1						3
Hogness, John University of Washington					1					2
Hornig, Donald Brown University	1	1	4							6
Hubbard, John University of Southern Calif.						1				1
Ikenberry, Stanley University of Illinois							1	3	1	5
Johnson, Howard Mass. Instit. of Technology	1	1								2
Jordan, Bryce The Pennsylvania State Univ.								1	1	2
Kennedy, Donald Stanford University							1	2		3

(continued)

Table 4.2. *continued*

President & Institution	70	71	72	73	74	75	76	77	78	79	80	81	82	83	84	85	Total
Lyman, Richard Stanford University					1												1
MacGrath, Peter University of Minnesota						2								1			3
McElroy, William University of Calif.-San Diego					1		1										2
Marston, Robert University of Florida											1	2	1	1			5
Moos, Malcolm University of Minnesota		1	1														2
Muller, Steven Johns Hopkins University							1	1	1	1							4
O'Connell, Stephen University of Florida	1																1
Odegaard, Charles University of Washington	1																1

Olson, James University of Missouri				1		1		1	1	5
Oswald, John The Pennsylvania State University		1								1
Posvar, Wesley University of Pittsburg	1			1	1		1	1		5
Ratchford, Brice University of Missouri			1							1
Rhodes, Frank Cornell University								1	2	3
Ryan, John Indiana University						1				1
Sanford, Terry Duke University				1	1		1	1		5
Saxon, David University of California						1	1	1	1	4
Shannon, Edgar University of Virginia	1									1
Shapiro, Harold University of Michigan					1					1

Table 4.2. *continued*

President & Institution	70	71	72	73	74	75	76	77	78	79	80	81	82	83	84	85	Total
Sovern, Michael Columbia University													2				2
Sproull, Robert University of Rochester										4		2		1			7
Swearer, Howard Brown University										1			1				2
Toll, John University of Maryland															1	1	2
Walton, Clarence Catholic University of America			1														1
Weber, Arnold University of Colorado											1						1

Wharton, Clifton Michigan State University				1	2	1						4
Wiesner, Jerome Mass. Instit. of Technology			1	1	1	1	1	1				6
Wyatt, Joe Vanderbilt University									2	1		3
Young, Charles Univ. of Calif.- Los Angeles						1				1		2
Zumberge, James University of Southern Calif.										1		1
4	6	6	4	13	11	9	9	6	13	5	12	23	21	19	12	173

Table 4.3. Congressional Committees before which Presidents Testified

Congressional Body	Instances of Presidential Testimony*
House	
Committee** on Appropriations	9
Committee on Budget	5
Committee on Education & Labor	11
Select Subcommittee on Education	13
Subcommittee on Postsecondary	24
Committee on Foreign Affairs	3
Committee on Science & Astronautics	5
Committee on Science & Technology	14
Subcommittee on Science, Research and Technology	8
Committee on Ways and Means	5
Other***	15
Joint	4
SUBTOTAL	116
Senate	
Committee on Banking, Housing and Urban Affairs	3
Committee on Commerce, Science and Transportation	7
Committee on Finance	4
Committee on Foreign Relations	8
Committee on Governmental Affairs	3
Committee on Labor & Human Resources	7
Committee on Labor & Public Welfare	13
Other***	12
SUBTOTAL	57
TOTAL	173

*Total number of presidents offering 173 instances of testimony is 61.
**Unless there were more than 5 instances of testimony before a subcommittee, the testimony was not listed separately, but included under the broader committee.
***Only 1 or 2 instances of testimony were provided for the committees included in this category.

stances, while they appeared before the Senate on 57 occasions (33 percent). When they addressed the House, the presidents were most likely to appear before the Committee on Education and Labor and its several subcommittees, especially the Subcommittee on Postsecondary Education. The presidents spoke to this Committee 48 times, offering 41 percent of their total testimony to the House before it. The other committee that presidents were likely to appear before in the House was the Committee on Science and Technology or its Subcommittee on Science, Research and Technology. The presidents addressed this committee 22 times (19 percent). These two committees account for 60 percent of the presidents' testimony before the House.

In the Senate, testimony was distributed more evenly between committees. However, the Committee on Labor and Public Welfare, which includes responsibility for postsecondary education, led with thirteen (22 percent) instances. The Committees on Foreign Relations, Commerce, Science and Transportation, and Labor and Human Resources account for roughly the same numbers of instances of testimony, together representing 22 instances or 38 percent of the testimony before the Senate. The Committee of Foreign Relations included consideration of international economic competition and international science issues, the Committee on Commerce, Science and Transportation looked at science policy issues, and the Committee on Labor and Human Resources touched broad educational policy issues, especially as these affected labor force problems. These four Committees—Labor and Public Welfare, Foreign Relations, Commerce, Science and Transportation, Labor and Human Resources—account for approximately 60 percent of the testimony before the Senate.

Although this data set represents policy positions espoused by AAU presidents in the federal policy arena, it has a number of limitations. First, it treats only the legislative branch of government, a serious limitation given that scholars have long thought the scientific community, if not the university community as a whole, was most effectively able to influence policy through the executive branch. Second, it says nothing about the way in which presidents arrived at the positions they present. The process of negotiation, the politics of writing policy positions, the constraints imposed by attempting to craft arguments able to influence legislators, all of whom have agendas of their own—all this escapes the data set under consideration. Third, the format and intent of Congressional testimony frequently compels presidents to justify requests for continued or increased monies. The very nature of the hearings, then, may markedly narrow the policy issues which presidents addressed.

Topics and Themes

The topics which preoccupied presidents were research (54 percent of all testimony) and higher education appropriations, particularly for student financial aid (19 percent of all testimony; see Table 4.4). Themes were coded only for these two major topics. The remaining topics did not provide a sufficient amount or depth of material to repay coding. They are treated simply as topics, in terms of the issues raised.

When testifying on the topic of research, presidents addressed general concerns that they faced as leaders of research institutions—research infrastructure, the National Science Foundation (NSF) budget, the interaction between university research and industry, between university research and the various mission agencies. These concerns

Table 4.4. Topics on which AAU Presidents Spoke

Topic	Instances of Testimony*	
Research	= 94	(54%)
General, Physical Science 64		
Special, Physical Science 17		
Other Than 'Hard' Science 13		
Student Aid: Higher Education Appropriations	= 33	(19'%)
Tax Law	= 14	(8%)
Intellectual Liberty	= 7	(4%)
Extraneous**	= 7	(4%)
Social Concerns***	= 6	(3.5%)
Government Regulations	= 6	(3.5%)
Student Legislation	= 3	(2%)
Miscellaneous Higher Education	= 3	(2%)
	173	100%

*The number of instances of testimony and the number of presidents who testified are the same. Each instance of presidential testimony was coded as one topic. When a number of presidents spoke at one session of Congress on the same topic, each testimony was coded as a separate instance.
**These presidents were called upon to speak because of expertise in areas not within the purview of their role as leaders of higher education.
***These concerns were when presidents spoke as individuals, rather than representatives of their institutions or the associations. The subjects were: ending hostilities in Indochina, enviromental protection, arms control, anti-apartheid legislation.

were almost always phrased in terms of the physical sciences. The mission agency hearings at which they appeared were, in order of frequency: NSF, National Institute of Health, Department of Agriculture, Department of Energy and Department of Defense. The presidents spoke too about special research projects housed at or related to their institutions—the National Laser Users Facility, the supercomputer, the Midwest Research Laboratories and the like.

The next topics to which presidents devoted a good deal of attention were student aid and general higher educational appropriations. These topics were addressed in 33 (19 percent) instances. Generally, this testimony was geared to fending off cutbacks in student aid.

The remaining 25 percent of presidential testimony covered a wide range of topics (see Table 4.4): tax law, the intellectual liberties of students and faculty, government regulations, and legislation related to students. None of these topics accounted for more than 8 percent of the total testimony.

By its very nature, the AAU presidents' Congressional testimony is a quest for resources. The federal government supplies a large part of the presidents' budgets, especially the research monies so central to the elite status of these institutions. Since World War II, the federal government has provided between 65 and 80 percent of their Research and Development (R&D) funding. The presidents are quite frank about the centrality of resources, and offer an unabashedly materialistic interpretation of scientific excellence. In the words of Jerome Wiesner, President of Massachusetts Institute of Technology (MIT), and former science advisor to several U.S. Presidents, "Some of the problems [of academic science] cannot be solved by money, but there are few if any which would not be easier to resolve if more were available."[8]

In their quest for resources for research and appropriations for higher education, four main themes emerge. First is the AAU presidents' efforts to preserve the elite status of their institutions and the existing structure of prestige in the scientific community. The second theme is national security. Four aspects of the national security theme reoccur; distress over the breakdown of the funding relationship between the Department of Defense (DOD) and research universities in the wake of the Vietnam War; conviction that basic research plays a central role in maintaining national security; ambivalence toward increased military funding; and promotion of a relationship between national security and economic prosperity that hinges on the development of high technology. These themes are contradictory, reflecting presidents' struggle to reconstitute a stable and predictable funding pattern after the Vietnam War. The themes are never fully reconciled.

The third theme is the development of a close alliance between these presidents and like-minded CEOs of large corporations. This alliance involves a gradual turning from an ideology of pure science to one that calls attention to the utility of basic research in high technology.[9] The emerging partnership between universities and industry also involves the shared articulation of a strategy for economic development likely to have sweeping consequences for the structure of work in the United States.

The fourth theme is the preservation of access for poor, minority and middle income students. Presidents regularly requested monies and programs for these students. Generally, they argued that higher education, especially private higher education, should not be the exclusive preserve of the children of the wealthy.

Theme: Preservation of Elite Status

From World War II until 1968, U.S. science enjoyed unprecedented growth, funded almost exclusively by federal monies. From 1955 to 1968, R&D expenditures rose from slightly over fifteen billion dollars to just under 30; basic research, most of which was performed in universities, rose from five billion to just under 20. Between 1968 and 1978, budgets declined slightly, and the value of science as a social undertaking was widely questioned.[10] The decline deeply disturbed AAU university presidents, who were accustomed to constant and rapid growth. The foundation of their institutions' prestige rested on the strength of the research enterprise.

After 1968, AAU presidents tried to rebuild their research capacity and preserve their elite status, even at the expense of other postsecondary institutions. Eight presidents gave 12 instances of testimony on the topic of research that developed the theme of preserving the elite status of their institutions (see Table 4.5). They sought to protect their position in two quite different ways. First, they advocated a reconcentration of funding on selective, high quality graduate programs. Second, they called for a restructuring of the presidential science advisory apparatus to provide expertise routinely from elite universities. The majority of this testimony occurred between 1973 and 1978.

In 1973, after efforts to stabilize higher education resources through institutional aid failed,[11] AAU presidents began to advocate concentrating federal resources on centers of excellence. In the main they argued that concerns about access should be confined to undergraduate school, while graduate education should be firmly based on merit. They worried about tendencies toward "egalitarianism," especially prevalent at state universities, and called for a greater emphasis on quality. Dale Corson, President of Cornell, put it most clearly. Speaking for the AAU in 1975, he noted that the Association had advocated institutional aid, but now recommended turning down annual institutional grants of $200 for each fulltime graduate student enrolled. He pointed to the scarcity of resources, and reasoned that:

> it is no longer serving the national interest to scatter that money all around the country. It has to go places that are producing the highest quality people and the best programs.... There are a limited number of institutions that are going to provide the leaders, and I believe that the national interest is best served by supporting that highly capable group of institutions.[12]

Of course, the capable institutions were by and large AAU universities, which, prior to the expansion of the postsecondary system, had received the lion's share of federal monies. Until 1965, approximately 80 percent of federal R&D funds were concentrated on the top 50 performers, which were almost coextensive with AAU institutions.[13] Corson advocated reconcentrating resources on AAU universities, explicitly arguing for discontinuance of programs developed in the mid 1960s by the NSF and DOD to improve institutions in the "second tier" and to broaden the geographic distribution of research funds.

The AAU took a similar position in a document entitled *A Report of the Association of Graduate Schools to the AAU,* which Corson entered into his testimony.[14] The AAU saw the expansionary period of the 1960s as past: the 1970s would be different. "Over the coming decade, continuing elevation of the level of excellence must be the primary objective." The AAU made a principled argument for "quality" and "excellence" even if "new" students—women, adults, part-timers—were not served: "One paramount principle is that areas of excellence in existing doctoral progams should not be sacrificed."[15]

As part of their efforts to maintain their status, AAU presidents tried to repair the damage done to the scientific advisory structure under Nixon. In 1972, Richard Nixon eliminated the Office of Science and Technology from the Executive Office of the President, and transferred it to the Director of the NSF. Nixon also did away with the Office of Science Advisor to the President, and the President's Science Advisory Committee (PSAC). He almost seemed to be punishing elite universities for the part played by fractious students and faculty in the storms of protest that had swept the nation's campuses over his handling of the Vietnam War. He certainly was trying to reprimand university scientists for what he and his staff perceived as their criticism of the war and its weaponry. Criticism from PSAC seemed especially disloyal to the Nixon administration since substantial R&D sums had been poured into universities that housed PSAC members, yet neither the weapons nor the social control techniques needed to bring about an American victory in Vietnam had been forthcoming.[16] As Jerome B. Wiesner, President of MIT, and himself a PSAC member, put it when describing the demise of that body before Congress: "blunt judgments from PSAC regarding the weapons systems, the validity of the Vietnam operations assistance and the policies on the ABM were not particularly welcome."[17]

The elimination of the science advisory apparatus was a serious blow to elite universities. From World War II onward, leaders of the AAU universities had staffed these offices, establishing a close relationship with a series of the nation's Presidents, and maintaining secure ties with the executive branch. This had enabled them to influence science policy in a protected environment, where they were not subjected to the political volatility of Congress, and was in part responsible for the always rising science budgets for elite universities that had characterized the first two decades after World War II.[18]

In the mid-1970s, AAU presidents, especially those who had held past positions in the now dismantled scientific advisory apparatus, sought to reinstate their influence. In his testimony before Congress over the organization of science and technology, Jerome B. Wiesner argued for the creation of an office whose primary allegiance would be to the President, one that would expand the availability of information accessible to him, free him from total dependence on the advice of the several government departments, and resolve conflicts between agencies. Donald Hornig, President of Brown, a member of the President's Science Advisory Committee under Eisenhower and Kennedy, and Johnson's Science Advisor, argued for creating a mechanism that would enable the President to get scientific input when problems were being formulated. William D. McElroy, Chancellor of the University of

California, a member of PSAC under Kennedy and Johnson, and a former Director of NSF, asked for the creation of science advisory post with Cabinet level status.[19]

In short, these university presidents were asking for a return to the organizational structure that had served them so well in the years before Vietnam turned into a quagmire. They had to wait until Nixon left office, and Ford instituted the Office of Technology Assessment (OTA). The men who had served on PSAC were never happy with OTA. Wiesner, who became the first chair of the OTA Advisory Council, always claimed that the Office was unable to handle the "tough problems."[20]

Theme: National Security and the Vietnam War

Despite AAU leaders best efforts to preserve their elite status, science funding in the mid 1970s remained stagnant or fell farther. Above all, AAU presidents sought stability and predictability in terms of resource flow. As Richard Lyman, President of Stanford, put it in 1974, "Our traditional method of handling both graduate education and research has been to provide a burst of support in reaction to a national crisis—to the Cold War, to the health crisis, to Sputnik, to the environmental crisis . . . " These spurts of effort, he said, were "very difficult to do without a steady level of support."[21] In their search for reliable sources of support, these men began to devote attention to analyzing their failure to maintain and increase research monies. Six presidents offered eight instances of testimony that had a thematic concern about the impact of the Vietnam War on traditional patterns of research sponsorship (see Table 4.5).

At a 1970 Congressional hearing devoted to Post [Vietnam] War Economic Conversion, Howard Johnson, President of MIT, outlined the model of science and technology transfer that had dominated university research until the mid 1960s. In this model, national security was of paramount concern, and technology transfer as well as industrial innovation followed as spin-offs. "The Department of Defense, the Atomic Energy Commission, and the National Aeronautics and Space Administration have been the largest supporters of campus research and development at MIT," he said.[22] Defense work led to development of products in the fields of communications, computers, inertial guidance, numerical control, structural design, cryogenics.

Table 4.5. Presidential Themes

Theme	Instances of testimony*	Number of presidents who testified
Preservation of elite status	12	8
National Security	31	13
and the Vietnam War	8	6**
and increased appropriations	8	6
and ambivalence toward DOD	6	5
and economic prosperity	9	7
Business-Higher Education Partnerships	25	16
Preservation of Access	22	15
Minority Students	18	11

*Instances of testimony do not match the number of presidents who testified (Table 4.3) because any number of themes could be coded per testimony, some presidents testified on more than one theme, and not all presidents addressed major themes.
**The number of presidents who testified on sub-themes does not match the number who testified altogether on National Security because the presidents speaking to the sub-themes often testified more than once.

> "We can trace well over $10 billion of annual industrial sales based on technical products that were invented or developed at MIT since the war [World War II].... The entrepreneurs who left MIT, spinning off from MIT research laboratories, have formed over 250 companies since the war."[23]

While Johnson was able to describe the way MIT had balanced national security and economic prosperity in the years after World War II, he could not fathom what had gone wrong, and was only able to defend the model that had served MIT well.[24]

The presidents sometimes seemed so taken aback by the breakdown in the post-World War II model of science funding that they spoke more to their difficulties in adjusting to change than to the problem at hand. For example, Jerome Wiesner, after invoking the many indicators of decline in the U.S. world science position, complained that maintaining a competitive scientific advantage was very trying. "One of the prices of leadership is that it's hard, you know. It's harder to be first and to stay first than it is to be second or third where you can do a lot of following."[25]

Some presidents blamed the student activism of the 1960s for the problems that the university had in the next decade. In 1979, A. Bartlett Giametti held the student "revolution" generally accountable for problems encountered by the university. He saw what he regarded as the excesses of the 1960s portrayed most clearly by the betrayal of language, which in turn undermined the university by creating bad thinking, a loss of discipline, a preoccupation with feeling, or sentimentality, and a loss of the notion that we should work hard and submit ourselves to the rational and precise discipline embodied by language and the basic curriculum once central to elite colleges.[26] Wesley W. Posvar, Chancellor of the University of Pittsburgh, speaking in 1977, was even more direct in blaming the troubles of the sixties for funding problems in the 1970s:

> We have been through some very difficult times in higher education in our defense of technology and research.... As we all know, because of the criticism of the past years, in constant dollars the Nation's basic research investment has dropped 20 percent, along with the NSF budget.[27]

Steven Muller, President of Johns Hopkins University, offered a somewhat different analysis. He saw the social and student movements of the 1960s not as a central source of university funding problems, but as a critical response to the university's over-dependence on the federal government. In clarifying this point, he offered an insightful analysis of the relation between the American research university and the government. He saw the relationship as held together by a shared commitment to national defense.

> In an overall sense the American university was mobilized for war by the federal government in 1941, and demobilization did not occur until 25 years later.... War's end [in 1941] brought not demobilization, but the Cold War. National defense research continued, still under university auspices, as the era of nuclear weapons, unmanned missles, and space flight unfolded. After Sputnik, federal sponsorship of research in basic science became for a time a national security priority.... The Cold War enveloped the Third World. Here also the universities enlisted. Foreign language and area studies under the National Defense Education Act; assistance to less developed nations by faculties from the Agency for International Development; and even training of Peace Corps volunteers by university staff, on and off campus—all were evidence of the continued mobilization of the American university in

the national security interest. So was the continued involvement of the university community with the national intelligence establishment, which began but did not end with the Office of Strategic Services in World War II. In this very general sense, the still mobilized American university was drawn into Vietnam as the U.S. became more and more deeply committed there. The disaster of that commitment troubled the partnership between government and university; in large part it produced that student revolt of the late 1960s; and ended in the partial demobilization of the 1970s....[28]

When Muller said that "federal sponsorship of research in basic science for a time became a national security priority" he touched on one of the major difficulties the AAU presidents had in developing new ways of thinking about how universities should be funded. Contradictory as it seems at first glance, the ideology of basic research or pure science was tightly linked to research for national defense. The DOD had managed research in ways compatible with scientists' desires for professional autonomy. The key to mutual approval was the merit system, peer review, and long term project grants. The presidents thought, as Johnson of MIT put it, "the Department of Defense agencies that supported research and development had exceptional insight and sensitivity into the way university scientists and engineers work."[29] David Saxon, President of the University of California, saw the Office of Navel research "as one of the most successful models of agency-university relations in our national experience...." What was good about the ONR program was the broad range of subjects, the long term view, the willingness to wait for years for a pay-off, a staff composed of scientists more concerned with science than the institutional advancement of an agency, and program officers who had the authority to act on their own. "The whole orientation of the agency was directed toward serving research programs and needs and finding shortcuts rather than following arbitrary procedures."[30]

What the presidents found so attractive in the way national security agencies handled research was that they gave the universities a great deal of autonomy. DOD and ONR let professors act as independent professionals: the setting of research goals was shared, little was asked in the way of accountability, few if any regulatory restrictions were imposed, there were no affirmative action constraints on personnel decisions. In Jerome Wiesner's words, DOD and the Atomic Energy Commission, which he considered a defense agency, were successful because the applied research in which they were engaged was done in the same ways that "we do basic research."[31]

The presidents who had participated in DOD research after World War II emphasized support for basic research, but probably fully understood the contribution research made to the development of weaponry and military systems. Unlike later generations in the academic community, these presidents saw no conflict between the needs of science and national security. Indeed, as we saw in Chapter 3, a good number of the presidents served as officers in the military. As Steven Muller, President of Johns Hopkins noted, the university was not coerced into partnership with the military: "Quite the contrary: mobilization for the most part reflected a partnership entered into with enthusiam, first at the outbreak of a war that was nearly unanimously regarded as just, and later sustained by the prevailing preference of both parties."[32]

Ironically, DOD monies helped to perpetuate an ideology of pure science. AAU presidents found it very difficult to give up an ideology that had created so much professional space for research. However, the Vietnam War, student revolt, Mansfield Amendment and Watergate weakened the ties between AAU universities and the defense mission agencies. Invocation of the need for national security was no longer enough to continue research funding. The presidents had to look for another rationale for a stable resource base.

Theme: Business–Higher Education Partnership

The new rationale AAU leaders developed in the late 1970s and the 1980s centered on the contribution academic science makes to economic prosperity, and focused on the practical utility of basic research. Between 1977 and 1985, 25 instances of testimony were given by 16 presidents that had as a major theme the ways that government could foster industry-university partnerships to create economic prosperity (see Table 4.5). The majority of this testimony was given after 1977.

Wesley W. Posvar, University of Pittsburgh, was the first of the presidents to speak directly about the utility of basic research to the business community. He argued that "basic research is important to economic growth in the solution of national problems...basic research is the key multiplier to economic development and in turn, to the development of our whole society."[33] Posvar was well aware that in linking basic research to economic prosperity, he was making a strong case for increased funding. He saw an emphasis on the contribution of applied research to economic recovery and the capture of global markets as more likely to stimulate Congress to invest in scientific infrastructure than abstract appeals for more money for basic research. He hoped university presidents as a group "had begun to learn that we have to explain ourselves and our needs better, and higher education and research are indeed part of the national interest, and not just the self interest of scholars . . . "[34] Although the presidents began to speak to the commercial advantages of fundamental research, they were always very clear that the federal government had to bear most of the costs of academic science. As Posvar said,

> normal economic incentives will not in themselves produce adequate volume of basic research . . . the lag between discovery and application is so long and unpredictable that only Government in our society can feasibly provide the necessary support. In other words, the horizons of industry are inherently too short to make this kind of investment . . . of course, the Federal agencies . . . do benefit.[35]

In effect, presidents were asking for a joint federal subsidy, one that would serve business as well as the university. This commitment to a common purpose made AAU universities and the leaders of powerful corporations, such as those represented in the Business–Higher Education Forum, political allies in pressing their claims on Congress. In these partnerships, the ideology of the utility of basic research was elaborated. Initially, the argument went as follows: although outcomes of basic research are difficult to predict, basic research must continued to be well funded because most technological breakthroughs stem from such research, unpredictable as it might be. In the words of Stanley Ikenberry, President of the University of Illinois:

> Basic research aims at the creation of new knowledge and seeks to understand the fundamental nature of processes. Although it is done without commercial application in mind, it is utterly essential to our economic progress and national security and, in the long term, frequently has remarkable practical application.[36]

The argument was elaborated to accommodate the social changes that business-university partnerships were creating, which in turn brought about a more rapid rate of technology transfer, and began to collapse farther the distinctions between basic and applied research. Donald Kennedy, President of Stanford, captured some of these dynamics in his 1983 testimony. He thought four forces had blurred the distinction between basic and applied. First, there was a genuine public concern about the rate of technology transfer, and the need to make it go faster. Second, new areas of science, from biotechnology to microelectronics, were so vigorous and exciting that they suggested their own applications. Third, universities had experienced a 43 percent real decline in non-defense federal support over the past fifteen years, and were looking for alternative funding sources. Fourth, changes in the economic structures supporting R&D contributed to the blurring of distinctions between basic and applied. These changes were characterized by:

> a new style of capitalization of high technology ventures. The new tax law changes four or five years ago have spawned a variety of new, high technology companies in which rapid and dramatic changes of value have been associated with the early possession of an idea.
>
> When a kind of financial incentive exists, you are going to see a migration of corporate interest to the earliest stages of this trajectory of innovation, because it is perceived that great changes in the value of corporate venture can take place if they are seen to possess important ideas. They do not even have to have products yet.[37]

University presidents were cognizant of the climate being created for investment in knowledge by changes in the law, by increased public support for high technology, and by the possibility of large, speculative profits.

The presidents saw university-industry research partnerships as an alternative source of independent funds for their institutions if they could make legitimate claims on the intellectual property produced. To make these claims, university presidents worked toward attaining a share of the profits from scientific discoveries. As John Toll, President of the University of Maryland, noted in 1984, the patent act of 1980 had cleared the way for university exploitation of patents applied for by faculty. The presidents, along with corporate leaders, wanted to expand the 1980 act so that it covered not only small business, but all business, including large corporations, allowed the university clear ownership of inventions, and removed the five year cap on exclusive licenses by universities to industry.[38] Universities would then benefit from a steady stream of royalties when business-higher education partnerships moved to production.

While the presidents hoped to achieve independence through royalties, they also wanted direct government subsidy of university-industry high technology partnerships. Direct subsidies called for a particular economic development strategy. In the works of Harold Brown, president of the California Institute of Technology, that strategy rested on "high technology utilization and high capital intensitivity."[39]

One of the manifestations of this development strategy was the "factory of the future." Presidents saw the university as participating in the creation of the "factory of the future" or "the automated factory." The possibility of such a factory has tantalized business leaders since World War II, after which time they periodically but prematurely announced its imminent arrival. The most salient feature of the factory of the future is that it has no workers. Joe B. Wyatt, Chancellor of Vanderbilt, perhaps most clearly voiced the AAU presidents' view of the factory of the future. He thought the demise of the conventional factory, which had depended on mass production, was signaled by prototypes developed at MIT which made automated, batch production possible by utilizing the computer and the machine tool. Computer Assisted Design/Computer Assisted Manufacturing (CAD/CAM) led to Flexible Manufacturing Systems (FMS), which were essential to dominating global markets. Automated plants, asserted Wyatt, were fifteen times more profitable, 30 times more productive, and required ten times less people to operate than the manufacturing techniques of a decade earlier. He saw history repeating itself: as the work force was displaced from farm to factory during the first industrial revolution, now it would be displaced from the conventional factory to the factory of the future.[40]

After 1980, in creating the work place of the future, university presidents and CEOs expected the state to serve as a silent partner, funding joint ventures between industry and universities that depended on heavy government subsidy. An example of the way in which these ventures work is provided by Richard Cyert's account of the Robotics Institute at Carnegie Mellon. The Institute had two major partners, Westinghouse, and the Office of Naval Research. In concert with their partners, the Institute was developing underwater robots. Westinghouse expected its initial investment of $1.5 million to be returned three times within the year. The enterprise was treated as a joint venture: "We meet quarterly to go over our projects, just as businesses might meet, and we are trying to estimate savings and gains from these projects and they look good."[41] No justifications were given for the healthy profit that Westinghouse would make using government funds; no mention was made of special monitoring mechanisms or pay back provisions. No exception was taken to a partnership between a defense agency, multinational corporation, and the university. Most testimony repeated this pattern: healthy corporate profits were made on government subsidized research at universities that were themselves recipients of a great deal of federal money.

In the mid 1970s, the AAU presidents began to request monies for research that would result in rapid development of high technology commercial products and processes. They justified these requests in terms of contributing to the U.S. struggle to maintain an edge in competitive global markets. In seeking this funding, presidents sought to create a climate in which it was possible for corporate and university leaders to act freely to realize profits from commercial research and development, jointly but unequally subsidized by the state, universities and corporations.

Themes: National Security—Increasing Appropriations, Ambivalence Toward Defense Spending, Linkage Between National Security and Economic Prosperity

When research budgets fell short of pre-Vietnam highs, the presidents cautiously began requesting increased defense appropriations for academic research. Although six presidents requested more monies in eight instances, five revealed a real ambivalence toward substantially larger defense appropriations (see Table 4.5). Toward the end of the period in question, presidents tried to reconcile their reservations about accepting more military monies by speaking generally to the role defense spending played in sustaining economic prosperity, a case 7 presidents made in 9 instances of testimony (see Table 4.5).

Although AAU leaders had begun to forge an alliance with business leaders, some still longed for what David Saxon remembered as "the halcyon days" of heavy DOD spending on academic science.[42] In 1979, Saxon, who was a member of the Buchsbaum Working Group at the DOD, asked the Senate Subcommittee on Science, Technology and Space for more DOD monies for research. He was speaking in his role as chair of the Committee on Science and Research of the AAU. He thought the Mansfield Amendment was unfortunate, believed that the DOD understood the requirements of basic research, and looked forward to a funding era in which the DOD would support those scientists whose peers thought they were doing outstanding or innovative work in broad disciplinary areas. The Working Group report "was addressed mainly to means of rebuilding and improving the basic research program of DOD and the generally good relations that the agency had enjoyed with the scientific community in and out of universities over the years."[43]

The Working Group exemplified DOD strategies aimed at recapturing university allegiances after the Vietnam War. This was made possible in 1979 when the Carter administration reinterpreted the Mansfield Amendment so that universities could again participate in almost all kinds of military research.[44] In essence, the DOD argued that third generation nuclear weapons depended upon processes being discovered by academic science in fusion, lasers, materials science, and artificial intelligence. According to the DOD, most basic research was defense research. Beginning in the late 1970s, the DOD offered substantial increases in research funds to the university. Most of these funds were in the 6.3 or "advanced applied" category, which addressed development of the technology base. "The DOD nearly doubled the level of technology base R & D obligations in real terms to academia between 1976 and 1986."[45]

At the same time the DOD was expanding its commitment to defense related academic research, it increased its spending for corporate R & D. The two streams of funding were joined by a common interest in high technology. As all fundamental research was defense research, so all improvements in the U.S. industrial base strengthened our defense capacity. "Investment in innovative manufacturing process technology strengthens the nation's industrial base not only by enabling contractors to produce better defense products, but also by improving industry's potential for quickly increasing output in periods of sustained conflict."[46]

By the beginning of the 1980s, DOD once again held more resources for research than any other organization, public or private. It developed funding rationales that supported fundamental research and commercial high technology research. DOD dollars were increasingly available for a wide range of industry-university-government partnerships.

AAU presidents, perhaps wary after experiences with student protest against military funding of academic research during the Vietnam War, did not come forward to advocate increased DOD budgets or greater participation in DOD research in large numbers. However, six presidents offered eight instances of testimony calling for increased appropriations (see Table 4.5). And on those occasions when they did speak out for a greater share of DOD monies, they acted as representatives of the associations of research universities. For example, David Saxon opened his 1979 testimony requesting increased DOD spending on academic research by saying, "I am here this morning speaking in my role as chairman of the Committee on Science and Research of the Association of American Universities."[47] In 1981, Robert Sproull, former head of the DOD's Advance Research Projects Agency and then President of the University of Rochester, spoke on behalf of the AAU, ACE, American Society for Engineering Education, Association of Graduate Schools, National Association of State Universities and Land Grant Colleges, and Council of Graduate Schools of the United States. He asked for more DOD monies, arguing that "the increasingly sophisticated weapons systems being developed by DOD depend heavily on advances made possible by basic research." He thought that "fields that DOD should be supporting in a significant way include mathematics, materials science, electronics, oceanography, engineering, computer sciences, condensed matter and radiation physics, atmospheric sciences, chemistry, and some of the social and behavioral sciences."[48]

After 1982, the AAU presidents were more ardent in their pursuit of defense dollars. In 1982, Reagan made budget proposals for 1983 that shocked the academic community. He initially asked for deep cuts in academic science, and almost eliminated social science research. The AAU presidents were outraged. They thought the need for funding for academic science to keep America's competitive edge in cut-throat world markets had become apparent in the last years of the Carter administration, that modest increases rather than deep cuts were more likely in the future. As James Olson, President of the University of Missouri, said about Reagan's proposed 1983 budget, "we face virtually nothing except cuts from almost every agency and almost every program with the exception of the Department of Defense."[49] Even David Saxon, who was not averse to increased DOD monies for the university, saw the proposed budget cuts as beyond bearing:

> We need a principle of reliable support. We need some understanding of the desirable mix of fundamental research, applied science and development. We need a principle to guide us in determining the long and short term potential of research. We need a principle of balance among public and private agencies to sustain diversity of support as well as balance among performers. Finally, we need a principle that excellence prevails over political and geographic considerations.[50]

Perhaps in response to the proposed budget cuts, Saxon became even more forthright in his pursuit of defense dollars. In 1982, he appeared before Congress and said: "I am here to talk about the readiness of the university community, particularly research universities, to contribute to the national security through research and teaching," and went on to detail ways in which this could be done, including reinstating the Reserve Officer Training Corps (ROTC). When asked during questioning whether the nuclear freeze or the growing debate over the government role in El Salvador would cause students to protest universities accepting DOD monies, Saxon virtually promised that students would remain quiescent.[51]

Nonetheless, AAU presidents, even those who ardently sought military monies, remained ambivalent about defense spending. Five presidents offered six instances of testimony that spoke to the multiple concerns they had about overly heavy reliance on the DOD (see Table 4.5). They worried about the deployment of military money, how efficiently technology transferred from the military to the civilian economy, the effects of heavy defense spending on the economy, and the consequences for science. In 1983, Robert Sproull noted that the increases in the Reagan administration budgets did not necessarily serve science because so much was earmarked for hardware:

> it [the defense budget] increases 29 percent in what's called Program Package 6, the RDT&E budget. In fact, the basic research part of that, the 6.1 money, goes up only about eight percent. Most of the increase then is in the large weapons program and the advanced development and engineering development associated with them. We think this is a poor balance. The basic research ought to grow as well.[52]

Sproull also noted that defense spending did not necessarily lead to efficient transfer of technology to the civilian economy, and suggested greater direct spending on research that benefitted the civilian economy. He pointed out that the federal government was:

> very practiced in sponsoring work in the aerospace industry, where the Government, through some Government agency, is the principal customer. But for most of what makes the country competitive with Asian countries and European countries, civilian industry is going to be the user of research. Therefore, it is in everyone's interest to have graduate students, and even undergraduates at universities, involved in research projects joint with industry, government and universities.[53]

Robert Q. Marston, President of the University of Florida, made the same point more directly:

> The U.S. Government spends substantially less than the governments of other industrialized nations on projects to stimulate industrial development and growth. Indeed, more than two-thirds of the Federal R&D monies are allocated to defense and space projects, and thus make a smaller direct contribution to productivity than would expenditures specifically for commercial projects.[54]

Both Sproull and Marston were suggesting that a federal spending policy favoring defense at the expense of industry was a questionable funding strategy. Implicitly, they were at least considering the notion that America's competitive edge was dulled by building the military rather than the civilian economy.

However, criticism of defense budgets was difficult to maintain, given the DOD's interest in revitalizing its relationship with the university. In 1983, at the request of the Committee of the Armed Services, the Under Secretary of Defense created the DOD-University Forum advisory committee. This organization was seen as a means of restoring a "healthier and more vital relationship between DOD and the university community." The Forum enables the DOD and universities "to address together, in candid and constructive discussions, the range and mutual concerns and opportunities that will shape future research and education programs."[55] Membership is jointly drawn from the DOD and the three higher education associations which the presidents most often represented in their congressional testimony: the Association of American Universities, the National Association of State Universities, and Land Grant Colleges and the American Council on Education.[56] In essence, the DOD was offering to renew the university-military partnership that had broken down in the 1960s.

If presidential response to DOD overtures can be gauged by their congressional testimony, by the mid-1980s presidents were seeking increased funding in both defense and commercial R & D, not making any sort of definitive choice between the two. The seven presidents who addressed defense spending in 9 instances of testimony after 1982 seemed to think that building our economic capacity was as important as our military and saw a strong global economy as critical to maintaining a strong defense (see Table 4.5). Increased funding for university research was central to both national goals.

Although the presidents as a group were probably well aware that the Strategic Defense Initiative (SDI) was in the offing, none of them supported Star Wars directly, and at least one was against it. A number of them were physicists or engineers who held security clearances. They had been at the forefront of weapons development, and had handled defense contracts for years. Marvin Goldberger, President of Cal Tech, talked knowledgeably about SDI the year before Reagan announced the program. When testifying as a private citizen in favor of SALT II and greater arms control though verifiable deep cuts, he said in January 1982:

> The final thing we kept in mind is that there is no hope that technology will provide some brilliant defense against Soviet missiles.
> Laser battle stations sweeping the skies clear is Star Wars stuff.
> Indeed, by the year 2000 we could put up large lasers on a satellite that would have some capability on missiles, but the cost of a real system would probably be in the trillions, terribly vulnerable, and easily countermeasured.[57]

While the presidents did not lobby for SDI, even when the initial phase of the program promised billions of dollars for basic research, nor with the exception of Goldberger, did they speak against it. And after 1983, the greatest increases in DOD academic R & D came in the 6.3, "applied advanced" category, rather than the 6.1 and 6.2 categories which have traditionally sustained fundamental research in the university.

In the early 1970s, presidents did not ask congress for increases in DOD funds for academic research. In the late 1970s and early 1980s, as defense budgets increased dramatically for the first time since the Vietnam War, they began to request larger DOD appropriations for academic science. After 1982, where the Reagan administration recommended sharp cuts in all but defense R & D budgets, they pursued military funding more directly and vigorously. However, the presidents did not rely on defense dollars exclusively, but continued to point to the necessity of continued R & D monies for research relevant to commercial product development. Seven presidents made this case in 9 instances of testimony. In the main, they began to blur arguments for the two lines of funding:

> Since World War II we have relied principally on the nation's universities to conduct the scientific research which has underpinned the technological innovations on which our economy and national defense are based.... The erosion which we have allowed to occur in our in-

vestment in research and education has far-reaching consequences, dulling our competitive edge and endangering the technological advantage on which the national defense is based.[58]

On the whole, the presidents were not able to develop a very consistent position on defense appropriations for academic research. Although they realized that it was impossible to return to the pre-Vietnam partnership between academic science and the DOD, they longed for the stability and autonomy of funding with which that partnership had provided them. During the Vietnam War, they came to understand the risks entailed by depending too heavily on a single sponsor. As the U.S. began to lose its competitive edge in international markets, some presidents began to question the wisdom of using defense as the primary agency for funding basic research. Their criticism was grounded in concerns about what increases in military funding would do to commercial research and development and in fears that military funding would move academic science definitively from basic to applied research. However, the increasing availability of DOD funds under the Reagan administration made it difficult to sustain criticism. Instead, presidents developed a broad rationale for funding, one that brought national security and economic prosperity together under the umbrella of high technology. By the end of the period in question, most presidents seemed able to justify and support all increases for academic research and development, regardless of agency.

Theme: Preservation of Access for Poor Students and Minority Students

Next to requests for more research money, the presidents spoke most often for more monies for student aid. The theme that emerged most strongly in their testimony was a commitment to maintaining broad access to higher education so that it did not become the exclusive preserve of children of the wealthy. Fifteen presidents offered 22 instances of testimony that touched on this theme (see Table 4.5). They probably had their finest moments in making such appeals. Consider John Brademas, President of New York University, speaking before the House Education Committee, on which he had sat for twenty-two years:

> Ronald Reagan's proposals for higher education represent, in effect, a

declaration of war on colleges and universities, especially independent ones, on students from both low and middle income families.... I fear we will move toward a two tier system of higher education... with independent universities for the rich, and state or municipal colleges for everyone else.[59]

Or A. Bartlett Giametti, President of Yale, who said of proposed 1986 budget cuts that they would:

erode the steady movement we have made as a nation toward equal access to higher education... choices and options will be limited once again by the accident of family wealth.... To encourage education was considered essential to the public good of the Nation, and it is one of the most distinctive aspects of our development as a country... These funds represent the best of our national character. They have been the catalyst for monumental change.[60]

Giametti said the entire fabric of Yale had changed since Yale had adopted a "needs-blind" admission policy twenty years ago. The student body was much more diversified and talented: seventeen percent of the 1984 class came from minority groups, 60 percent from public secondary schools, and the level of academic ability had risen.[61]

However, pleas for more student aid were not without a modicum of institutional self interest. One third of the student aid requests were for institutional aid, where government monies would be given to colleges and universities rather than going with the students. Of the requests made after 1981, (55 percent of the total), 78 percent were made by the presidents of private universities. The choice that Brademas and Giametti sought for students meant providing enough monies for them to attend institutions whose total yearly costs were probably about $18,000 by 1986. The Reagan cuts hit private institutions very hard by striking at all programs except those for most needy. In effect, the presidents were asking for a greater subsidy for the private sector and the middle class.

Eleven presidents spoke on the theme of continued support for minorities and the poor in eighteen instances (see Table 4.5). Minorities and the poor were mentioned, not dwelt upon, and women were rarely spoken of as a discreet category. Very little sustained attention was given to ways in which higher participation by minorities could be achieved. The only time inequities were addressed was if they were particularly glaring, as in the case of the Graduate and Professional Opportunities Program (GPOP). The GPOP scholarships were designated for minorities and offered a $4500 stipend as compared to NSF and DOD stipends running at $8100, plus, in the case of DOD, a matching allowance for the institutions. Almost no attention was paid to the impact that the high technology economy and military development strategies being pursued by AAU presidents might have on women and minorities. Only John Brademas briefly addressed these issues. He noted that "women and minority students are concentrated in the fields hit hardest by the decline in federal funding, so precisely the students who most need ... aassistance are most adversely affected."[62]

Despite the lacuna of self interest in their testimony, the presidents spoke regularly for maintaining broad access, particularly for minorities. In terms of student aid, they stressed the centrality of merit based access for the higher education system as a whole. With regard to minorities and sometimes women, they spoke to the importance of representation in a multicultural society.

Topics: Tax Law, Classified Research and Government Regulation

The presidents spoke on a number of topics that did not occur with enough frequency to lend themselves to thematic analysis. Nonetheless, the topics are important. They are tax law, classified research and government regulation (see Table 4.4). Several topics are not treated because they are not relevant (extraneous, individual social concerns) or because they are diffuse and few in number (student legislation, miscellaneous higher education).

In terms of taxes, AAU presidents routinely supported regressive taxation patterns. There were fourteen instances of testimony with regard to taxes between 1973 and 1984 (see Table 4.4). Presidents asked for tuition tax credits, provided these were not in any way exchanged for direct aid measures. They were against tax reform if the charitable deduction were not retained at the highest marginal tax rate, or unless reductions in incentives to give were offset by alternative mechanisms that stimulated private support.

Seven of the presidents addressed the topic of intellectual liberty when they spoke to military or economic classification of academic research (see Table 4.4). All spoke strongly against it. They argued for "security by achievement" rather than "security through secrecy," phrases that went back to debates following World War II about how freely information should flow in academic science.[63] Indeed, they asserted that major universities would not accept classified projects.

Although the presidents began with strong assertions about the need for an absolutely open atmosphere on campus, they usually qualified that position considerably during the course of testimony. The nature of the research in which the university was engaged in the early 1980s made an absolutist position on openness untenable. Basic research was identified with high technology, an area where demands were made for economic and military security.

Industry-university partnerships often involved trade secrets, and patent applications, both of which depended on secrecy. As John S. Toll, President of the University of Maryland said: "If businesses are to make an investment at that stage [very early], they have to know in advance, before the discovery has been made, that their investment will lead to patents which will be protected during the period of development."[64] To accommodate the patent process, many AAU universities agreed to publication delays ranging from thirty days to one year. Long term contracts between universities and businesses, such as the Harvard-Hoescht contract, also called for company pre-publication review to assure that patentable ideas were covered and trade secrets kept.

Secrecy surrounding high technology commercial product development was heightened when the Reagan administration subsumed it under the general heading of East-West conflict. Under export control legislation administered by the Commerce Department, the selling of dual use technology to adversarial countries was prohibited, and sporadic attempts were made to stop the exchange of scientific knowledge that might lend itself to military ends. The federal government began to monitor scientific conferences and withheld permission for business projects in Soviet bloc countries. Both business leaders and university presidents fought against these restrictions. However, the blurring of distinctions between basic and applied research, as well as commercial and military made it difficult to draw clear lines, unless leaders were prepared to jettison a Cold War approach to global strategy, which they appeared not ready to do. Many of the technologies—such as super computers—were dual use. Moreover, the requests business had made for secrecy with regard to patents and trade secrets seem to have contributed to a climate that made government demands more plausible.[65] If secrecy were justifiable for multinational supremacy in global markets, why not for dual use technologies that might confer a cutting edge in military development?

Although the presidents spoke vehemently against secrecy, most conceded that some regulation was necessary. As Paul Gray, President of MIT put it: "The national interest lies not in either extreme, but in selecting goals and policies which serve both the cause of national security and the larger process of science and technology on which so much of this Nation's strength depends."[66]

The degree of regulation they were willing to participate in was perhaps best outlined by Dale Corson, President Emeritus of Cornell. He had been chair of the panel on secrecy put together by the National Research Council. The panel had developed the "gray area" concept, which tried to define a narrow area in which technology was rapidly developing, was dual use, was unobtainable by the Soviets from U.S. allies, and which would give the U.S.S.R. a near term advantage.[67] Universities and scientists engaged in work in the gray area would voluntarily withhold information from wide circulation.

However, the "gray area" proved difficult to define and scientists complained of self-censorship. Debate over regulation of information moved into the DOD-University Forum. Here, the presidents, in concert with DOD officials, drafted a position that, at first glance, buttressed academic freedom by bringing greater clarity to the issue:

> ... the mechanism of control of fundamental research in science and engineering ... is classification ... prior to the award of a ... grant or contract ... No restrictions may be placed upon the conduct or reporting of fundamental research that has not received national security classification.[68]

However, the caveat, providing for "periodically reviewing all research grants or contracts for potential classification," undermines the position initially taken.[69] If SDI is developed, the strategic importance of basic processes under investigation will make increased classification and classification after the fact more likely. University presidents will probably continue to resist encroachment in their domain, but will confine their arguments to organizations like the DOD-University Forum, where they are far from the public eye.

Six AAU presidents wanted to repeal regulatory measures, such as the anti-trust laws and affirmative action (see Table 4.4). In terms of anti-trust law, two presidents reasoned that prohibiting corporations from acting together placed American firms at a disadvantage in world markets. The West Europeans, and especially the Japanese, frequently formed pools or cartels, often subsidized by the government, which gave them a competitive edge in targeted, high technology industries. AAU presidents asked for similar opportunities for U.S. business. As Richard Cyert, President of Carnegie-Mellon, saw it, during the past 85 years the Sherman Act had been interpreted as inward looking, fostering competition within internal markets. Currently, what was called for was an outward looking policy, one that guaranteed competition for external markets and provided for more joint actions and investment pools between companies.[70]

AAU presidents were also begged relief from affirmative action regulations. Four testified at hearings specifically devoted to the subject, making the case that its costs were too high, and its intrusion into the internal workings of the university too great. They saw themselves as firm supporters of affirmative action in the abstract, but were unhappy with the concrete regulatory measures offered by the federal government. They wanted less red tape, less monitoring, more sensitivity to the unique characteristics of the university, an end to the adversarial posture taken by so many investigators, and a simplified procedure for data gathering.[71]

In essence, these presidents seemed to be asking for affirmative action regulation modeled after industry specific agencies, such as the Interstate Commerce Commission, the Federal Communications Commission and the like. They shared this position with business leaders, who found cross-industry, as opposed to industry specific, regulation a constraint on managerial initiative. Unlike cross-industry regulation, typified by areas such as affirmative action, environmental quality, occupational health and safety, industry specific regulatory agencies historically were so sensitive to the needs of their clients that the interests of agency and industry were often hard to separate.[72]

Similarities and Differences with the Business–Higher Education Forum

There are some direct connections between presidential testimony and the Business–Higher Education Forum. In at least three instances in the early 1980s, presidents refer to their work with the Forum. For example, in 1982, Wesley Posvar, Pittsburgh, testifying on the importance of university-industry partnerships, mentioned that he served as Chair of the Business–Higher Education Forum for two years, and that it was concerned with "forming a common alliance to deal with the kind of problems we are concerned with here this morning."[73] Robert Marston, president of the University of Florida spoke before the House as a representative of the Forum in 1983. Essentially, he requested more support from Congress for Business–Higher Education Forum policies, invoking increased economic competition from the Japanese and Europeans. He spoke most strongly for development of technology transfer between universities, government and industry.[74] In 1983, Posvar again mentioned his work with the Forum, stressing the "spontaneous initiative" behind business–higher education partnerships, which he saw as providing a model for meeting future research needs for higher education.[75]

More significant than these few instances of direct testimony are the underlying topical and thematic similarities between the Forum positions and the presidents. These similarities are especially striking during the later years of the presidents' testimony, when a number of them had worked with the Business–Higher Education Forum for several years. Indeed, a major theme for the presidents (see Table 4.4) from the late 1970s forward was development of business–higher education partnerships. Rather than being coerced or coopted into these partnerships, the presidents appear to have sought out and promoted them. They see the partnerships as an alternative to an overdependence on federal funds, such as occurred with heavy reliance on the DOD, and as a potential source of independent income through patent royalties.

As indicated in Chapter 2, the Forum had an initial legislative program that stressed four main points: (1) increased tax cuts, (2) rolling back anti-trust and (3) regulatory legislation while (4) developing intellectual property legislation. The AAU presidents touched on all of these, albeit sometimes briefly, and usually with regard to their relevance for higher education as a policy arena. They spoke most frequently to taxes (see Table 4.4), seeking tax breaks for universities. They advocated reinterpretation of the Sherman Anti-Trust Act, asking for greater corporate cooperation domestically, and promoting the possibility of pooling research, especially university-industry partnerships, among corporations in the same enterprise. They sought a substantial revision in affirmative action legislation, requesting special privileges for higher education on the basis of the expertise required to make judgments about personnel decisions. When speaking to industry-university partnerships, presidents championed intellectual property laws that would allow the university to hold rights to intellectual property discovered by faculty.

Although the presidents were almost always addressing university rather than corporate issues when they spoke for points central to the Business-Higher Education Forum legislative program, the congruence between university and corporate programs is notable. The presidents seemed to think that the strategies pursued by corporations serve universities well. At some levels, on some points, they see university and corporate interests as closely related.

There are many similarities between university programs and the Forum, but there are also differences. Perhaps the most notable differences are university presidents' interest in increasing student aid appropriations and in restraining efforts to classify scientific work. Although presidents comprise approximately half of the Forum membership, these interests are not strongly reflected. Forum interest in higher education is not general, but directed toward specific programs designed to enhance America's ability to compete in the global market. Forum interest in intellectual liberty is directed more toward easing government restraint of export of trade and technology than toward creating a open atmosphere for conducting science. These differences will be more fully explored in Chapters 5, 6, and 7.

Conclusion

The presidents' testimony enables us to consider some of the theoretical questions initially raised about access to policy and the dynamics of change.

Pluralist theory sees individuals as able to influence the policy process. If the individual cannot shape policy independently, then he or she can work through interest groups. During the period under discussion, 1970-1985, presidents acted more and more like members of an increasingly cohesive interest group. Testimony mounted in frequency. In the first five years (1970-1974) there were 36 instances of testimony, in the second (1975-1979) there were 50, in the third (1980-1984), 77. The number of presidents testifying on a single issue also increased, especially after 1980. Presidents seemed to act as a group representing certain higher education associations to Congress in response to budget cuts proposed by the Reagan administration. They did not limit their efforts to Congress, but worked also with the DOD and NSF, making multiple approaches to the state.

However, presidents did not come together simply as presidents. In their testimony, they emphasized again and again organizational affiliations that reinforced their legitimacy as representatives of the most prestigious segment of higher education, the research university sector. This behavior more nearly fits with elite pluralist interpretations, which recognize inequities of power within a sector.[76]

But elite pluralist theory, like neopluralist theory, expects all interest groups to have ready access to policy, unless the policy directly concerns business, in which case it would be anticipated that the corporate sector would have the strongest voice. AAU presidents were clearly dubious about their ability to influence policy acting independently, as evidenced by their readiness to develop and engage in various business and higher education partnerships. Their testimony indicated that they saw the corporate community as an increasingly necessary political ally.

A neo-Marxist interpreting this same pattern of participation might read the AAU presidents' approach to the corporate community as predictable, with presidents espousing business and higher education partnerships because they lack the power to realize their own agenda and are resource dependent. Neo-Marxist theory would probably see the presidents turn to corporations as dialectically impelled by the societal conflict in the 1960s, which resulted in some state withholding of resources. The interpretation overlooks two points. First, presidents did not slavishly reproduce a business agenda. Generally they engaged corporate agendas when these were congruent with their own needs, something that occurred frequently but hardly exclusively. Indeed, the presidents did not move toward business mechanically and inevitably, but at that time when there was a possibility that they might become partners in productive enterprises, from which they could reap institutional profits, independently controlled, raising the possibility of a more vigorous autonomy at a future date. Second, the presidents always defended critical sectorial interests, such as increased monies for student aid, even though corporations demonstrated no interest in policy of this nature. Presidents were not so much subservient to business interests as they were looking to articulate their interests with those of corporations, to the benefit of both. Whether this is a shared class consciousness or not is difficult to decide.

Corporatist theory would look at the policy activity of university presidents and corporate leaders through another lens. Panitch offers a minimal definition of corporatism as:

a political structure within advanced capitalism which integrates organized socioeconomic producer groups through a system of representation and cooperative mutual interaction at the leadership level and mobilization and social control at the mass level.[77]

At the leadership level, corporations are historically the central organized socioeconomic producer group. Research universities are central to production in two ways. First, they train and prepare the highly specialized personpower necessary to large scale production of high technology. Second, the university itself is now moving aggressively toward a producer position, engaging in all manner of entrepreneurial science, very often in partnership with the state and corporations. These peak associatons may have come together to represent the interests of their several sectors around production issues and to introduce shared views into the policy process at a time when interest group activity has fragmented the political process and made any widespread consensus on policy extremely difficult.

The greatest flaw in this interpretation is the lack of attention to the mass level, and especially the absence of any serious consideration of labor, essential to productive enterprise across sectors. Indeed, many of the policy positions taken by presidents are implicitly anti-labor. High technology is capital intensive, not labor intensive, designed to heighten the productivity of a smaller number of workers rather than increase the number of jobs overall. Weakening anti-trust legislation is more likely to strengthen the hand of corporations than to enhance the ability of workers to take collective action. State subsidies for industry-university partnerships bring no immediate returns to workers organized in traditional blue collar industries. Indeed, labor, with its wage demands, is seen as one of the root causes for the loss of our competitive edge in global markets.

In sum, each of the theories explains partially university patterns of participation in policy making, but none accounts fully for the patterns presented in this chapter. Each of these theories has a history of its own that, in turn, responds to concrete historical conditions. For scholars situated in the university, pluralism may have seemed an adequate explanation for policy making in the period of post war prosperity, but not up to the task of making sense out of the social and political upheavals surrounding the Vietnam War. At this point, some scholars began to turn to Marxian interpretations, which better dealt with inequalities of power, the relationship of the polity to the economy, and conflict. However, the tumult of the 1960s did not develop into class struggle, and capitalism did not collapse of its own contradictions. As neopluralists and neo-Marxians tried to deal with those aspects of history which their theories did not explain, corporatism, especially as a partial structure, began to seem like a way to account for the enduring and powerful interest associations that influence policy in liberal democracies. However, corporatism in the United States does not deem to be concerned with mass organizations and corporatism in general does not seem to deal well with the unpredictable but continued independence of mass organizations and social movements, all of which undoubtedly calls for new theoretical iterations.

CHAPTER 5

Bases of Corporate Interest in Higher Education

In the preceding chapter, the analysis focused on why, in theoretical terms, university presidents are forming partnerships with corporate leaders. In this chapter, the analysis centers on understanding why CEOs are working with university presidents in organizations such as the Business–Higher Education Forum. The Forum addresses the emerging relationships between the two sectors in its statement of purpose. The Forum will work "to identify, review and act on selected issues that relate to the current and future requirements of business and higher education ... and to help guide the evolution of relationships between corporations and institutions of higher learning, while preserving their separate historical functions."[1] This statement, however, does not address the basis for cooperation, which is the point at issue here.

The several theoretical traditions would explain cooperation between the two sectors on very different grounds. Pluralists would probably offer a functional explanation, in which corporations exchange resources for university research, with each participating in the relationship from positions of rough parity.[2] Neo-Marxian interpretations would probably see corporations, driven by the imperatives of capitalism, as compelling research universities, through various mediations, to serve production needs by developing entrepreneurial science. Corporatist theory would expect the partnership between the two groups to center on their relationship as representatives of socioeconomic producer groups. Their common interests in production would override major differences.

By far the largest literature that addresses the basis for cooperation between corporate and university leaders is in the functionalist vein. This literature is descriptive and pragmatic, documenting emerging partnerships between the corporate and academic community with an eye to improving them. A large part of this literature is generated by state and science policy bodies such as the National Science Board and the National Science Foundation.[3]

Although no clearly articulated theoretical framework informs these studies, there seems to be a number of buried functional assumptions which, when excavated, form a line of argument or working theory about the relationship of the university to industry. In general terms, the basis of cooperation between the university and corporations is explained as follows: the university is best able to perform basic research; basic research is related to stimulating technical innovation; innovative corporations create jobs and markets; small innovative corporations often initiate high technology production efforts and generate high-paying jobs able to replace increasingly unstable blue-collar occupations; large corporations need university research to increase their innovative, competitive edge; by pooling their efforts, corporations and universities benefit each other; the joint efforts of universities and corporations in high technology or science-based industry is the key to American success in global markets; and American economic success is critical to societal well-being.

In essence, this is a functional "research for resources" argument, in which both parties benefit. It is functional too in its assumptions about the way change occurs. Discovery and technology are the key to change. Large and small firms have an equal opportunity to benefit from such discovery. Rational policy formation should be predicated on negotiations over the best way to facilitate discovery. Although there might be conflicts of interest between groups over which policies best realize the improved production of commercial technology, there is a conviction that reaching this goal will enhance the public good.

Generally, critical and neo-Marxian arguments about the nature of the new relationship emerging between corporations and universities are less well developed than functionalist theories. They focus on the accelerated growth of a military-university-industrial complex. Corporations are the dominant fraction, but the several fractions share broad class interests derived from their mutual need to maintain and increase power and profits at the expense of other classes. The military-university-industrial complex is compelled by the logic of multinational capitalism, which depends on high technology to keep first world countries in a dominant position despite a growing global division of labor and increased capitalist production in third world countries.[4]

In the United States, corporatist theory has not yet addressed the basis of cooperation between corporations and universities. However, on the basis of my reading of corporatist theory thus far, cooperation would depend on a group holding sectorial status as a socioeconomic producer. In the case of the Business–Higher Education Forum, CEOs would represent the business sector and university presidents would represent entrepreneurial science, which includes technology transfer, incubators, industrial parks, and equity positions in firms headed by faculty. These groups would then develop organizational forums that would allow them to develop consensual positions on policy touching the commercial production of high technology

Methodology

Data about the corporations whose CEOs hold membership in the Business-Higher Education Forum were developed to see if they could shed any light on the basis of cooperation between CEOs and university presidents. I thought the organizational form, product interest or geographical location of the corporations might speak to the nature of the emerging relationship between corporations and universities, and I developed an inventory of the corporations in terms of type, size, product line, and geographical location of production sites and sales offices. Another line of data was developed to address the specific nature of the corporations' interest in universities: did the Forum corporations invest heavily in research, display a great deal of activity in university gift programs, or make extensive use of university expertise in high technology? Yet another set of data was examined to see if corporate leaders displayed the sense of social responsibility sometimes attributed to professional managers.[5] Could CEOs' participation in the Forum be explained by a heightened sense of social consciousness? To answer this question, I looked at nonfinancial data on the number of Department of Defense contracts CEOs firms held, their investment in South Africa, their citations for environmental and civil rights violations, and their involvement in National Labor Relations Board disputes. Finally, I looked at the remarks that Forum CEOs made in the general press for the period 1978–1985. These are the years in which the CEOs were active in the Forum; I thought they might speak directly to their reasons for participation in partnerships with universities. Details of data set construction are presented as the several samples are analyzed.

The Corporations

I selected 46 corporations for study (see Table 5.1). Each participated for two or more years in the Business-Higher Education Forum.[6] I obtained data on the corporations from a variety of standardized sources. I used the several *Moody's* manuals to develop information that described the corporations and their research activity.[7] In addition to *Moody's*, I also consulted the *Fortune* ratings on the corporations, the *Council for Financial Aid to Education Casebook*,[8] the *Concerned Investors Guide*,[9] and *Stocking the Arsenal: A Guide to the Nation's Top Military Contractors*.[10]

Of the 46 corporations that participated in the Business-Higher Education Forum, 40 are publicly held and listed in *Fortune* or one of the several *Moody's* manuals. Of the 40 ranked by *Fortune*, 28 (70 percent) are industrial, 3 (7 percent) are banks, 3 (7 percent) are utilities, and 2 are retail stores (5 percent). There was 1 transportation company, 1 life insurance company, 1 diversified financial company, and 1 diversified service company.

When ranked in terms of net sales, the corporations were high on the *500* list. If only industrials are considered, 18 (64 percent) are in the top 100, 7 (25 percent) are in the next 100, and 3 (11 percent) are scattered among the lower 300.[11] As a group, then, the corporations are among the most successful in the nation.

The *Moody's* manuals offer a slightly different classification than the Fortune 500, providing fewer categories. Forty of the 46 corporations were listed in *Moody's*. Of those 40, 31 (77 percent) are classified as industrials, 4 (10 percent) as bank and financial, 3 (7 percent) as public utilities, and 2 (5 percent) as transportation. As in the case of the 500, the great majority of the corporations are industrials engaged in manufacturing. This analysis will concentrate on the 31 corporations classified as industrials in *Moody's*. I focus on the industrials because the Business-Higher Education Forum's major thrust, detailed in Chapters 6 and 7, is on revitalizing American manufacturing. I used *Moody's* because of the much greater depth of information provided.

The majority of the industrials that appear in *Moody's* are incorporated in Delaware, because of that state's historic accommodation of corporations.[12] They are quite venerable. Eighteen (58 percent) were in-

Table 5.1. Corporations Participating in the Business-Higher Education Forum for Two or More Years and Their Representatives

Aetna Life Insurance: William O. Bailey
Air Products and Chemicals: Edward Donley
American Can: William S. Woodside
American Telephone and Telegraph: James E. Olson
AmSouth Bancorporation: John W. Woods
Armco: Harry Holidy, Jr.
Atlantic Richfield: W.F. Kieschnick; L.M. Cook
Carter Hawley Hale Stores: Philip M. Hawly
Coca Cola: Donald R. Keough
Consolidated Natural Gas: G. J. Tankersley
Emerson Electric: Charles F. Knight
Federated Department Stores: Donald J. Stone
First Bank System: George H. Dixon; Dewalt H. Ankeny, Jr.
Ford Motor: Philip Caldwell; Donald E. Peterson
Fuqua Industries: J.B. Fuqua
Gannet: Allen H. Neuharth
General Dynamics: David S. Lewis
General Electric: John F. Burlingame
General Motors: Howard H. Kehrl; Roger B. Smith
Greyhound: John W. Teets
Harris: Joseph A. Boyd
H.J. Heinz: A.J.F. O'Reilly
HNG Internorth: Samuel F. Segnar; Kenneth L. Lay
Honeywell: James J. Renier; Edson W. Spencer
Levi Strauss: Walter A. Hass, Jr.
Mercantile Bancorp: Donald E. Lasater
Microelectronics and Computer Technology Corporation: B. R. Inman
Millipore: D.V. d'Arbeloff
Minnesota Mining and Manufacturing: L.W. Lehr; Allen F. Jacobson
Mutual of Omaha Insurance: V.J. Skutt
Pfizer: Gerald Laubach
Potlach: Richard B. Madden
PPG Industries: L. Stanton Williams: Vincent A. Sarni
RCA: Thorton F. Bradshaw
Rockwell International: Robert Anderson; Donald Beall
Sante Fe Southern Pacific (formerly Southern Pacific): Alan C. Furth
Syntex: Albert Bowers
Teledyne: George A. Roberts
Textron: Robert P. Straetz
TRW: Ruben F. Mettler
U.S. Fidelity and Guaranty: Jack Mosely
United States Steel: Thomas C. Graham; David M. Roderick; William R. Roesch
U.S. West: Jack A. MacAllister
United Telecommunications: Paul H. Henson
Westinghouse Electric: Douglas D. Danforth; Robert E. Kirby
Whirlpool: Jack D. Sparks

corporated between 1890 and 1920. Five (16 percent) were incorporated between 1921 and 1950, and 8 (26 percent) between 1951 and the present.

The corporations manufacture, distribute, and sell a panoply of products, everything from soft drinks to engineered fasteners, from automobiles to steel. If a conglomerate is defined as a corporation in which no more than one product accounts for 70 percent of sales, and in which there are two or more products that do not share common inputs or bear a relationship to one another, then 11 (35 percent) are conglomerates.[13]

Although the majority of industrial corporations in the Forum focus most heavily on a major product line, many (64 percent) have high technology segments. These are frequently recent and are engaged in the manufacture of products and processes such as thermonuclear fission devices, satellite systems, extraterrestrial control and command systems, microelectronics, robotic systems, machine intelligence with research and development capacity, genetic engineering, biotechnology, lasers, and a wide variety of communication and information technologies.

Whether industrials or not, almost all of the corporations on which information was available are multinationals, if this is taken to mean simply that they have a substantial presence overseas. Almost all of the 31 industrials have international divisions and sales offices offshore. Of the 31 industrials, 26 (84 percent) have manufacturing plants outside of the United States.

Although wholly owned consolidated subsidiaries are less closely related to the structure of the parent organization than manufacturing plants, more detailed information on them is available. Table 5.2 shows the distribution of these subsidiaries and gives an idea of the multinational presence in this area. Half are in the United States, followed by Europe (20 percent), Latin America (9 percent), the second world (Canada, Australia, New Zealand, 7 percent), the third world (3 percent) and Japan (1 percent). The remaining 10 percent are subsidiaries for which no data on location were provided.

The most sophisticated multinationals are those which have organized their production globally, sourcing all phases of production for maximum economy and efficiency. At this level, raw materials can be extracted from any place in the world, and parts can be manufactured in the United States, assembled in Japan, packaged in Korea, and returned to the United States. In the United States, few companies have achieved this level of sophistication. Approximately 10 (32 percent) of the 31 industrial corporations in the Business-Higher Education Forum have this degree of multinational presence.[14] Eight of the 10 are deeply involved in high technology.

The Forum corporations do not seem to conform to the expectations of functionalist theory. They do not demonstrate heterogeneity. Small firms as well as large should appear in firms engaged in high technology, fulfilling functionalist predictions and confirming studies done in the early 1980s which pointed to the importance of small, dynamic high technology corporations in creating new jobs.[15] The majority of the firms participating in the Business-Higher Education Forum are the obverse of small and dynamic; instead, they are corporate behemoths that were organized during the consolidations that marked the end of the first phase of the industrial era.[16] If an exchange of research for resources is at the heart of corporate and university relations, the corporations in the sample should be characterized by their commitment to high technology. Although the corporations have a strong interest in the sorts of activities sponsored by the NSF and research universities, their concern with high technology cannot be considered a definitive characteristic of the group.

A neo-Marxian interpretation would expect Forum corporations to be established multinationals, intent on moving production overseas to lower labor costs and to contain the power of trade unionism. Control of high technology developed in universities would be an instrument for ensuring that offshore producers remained dependent on knowledge generated in the United States and did not too quickly become producers in their own right. This interpretation would explain the high technology segments of many of these firms as the cutting edge and future locus of control of world markets, but would expect profits to continue to come from traditional manufacturing made cost competitive by cheap labor in offshore plants. Corporate interests would necessarily dominate university research agendas, absorbing the talent of gifted scientists for commercial purposes. Although this interpretation explains the composition of the Forum corporations and their geographic deployment, it does not explain their rather limited involvement in research, whether in-house or in universities, addressed later in this chapter, nor does it deal with the universities' own interest in entering the productive process, touched upon in the previous chapter.

Corporatist theory would not see the composition of Forum corporations as problematic; it would expect the business sector to be represented by large and powerful corporations. Corporatist theory would

Table 5.2. Wholly Owned Consolidated Subsidiaries of Forum Corporations

Name	Subsidiaries	United States	Europe	Second World	Third World	Latin America	Japan
Aetna Life Insurance	0	0	0	0	0	0	0
Air Products and Chemicals	21	12	7	1	0	1	0
American Can	155	131	3	1	0	19	1
American Telephone and Telegraph	0	0	0	0	0	0	0
Amsouth Bancorporation	4	0	0	0	0	0	0
Armco	26	9	6	3	0	8	0
Atlantic Richfield	2	0	0	0	0	0	0
Carter Hawley Hale Stores	20	18	0	2	0	0	0
Coca Cola	29	17	3	1	1	6	1
Consolidated Natural Gas	13	13	0	0	0	0	0
Emerson Electric	121	33	50	13	5	15	5
Federated Department Stores	1	0	0	0	0	0	0
First Bank System	74	0	0	0	0	0	0
Ford Motor	23	4	10	3	1	4	1
Fuqua Industries	29	24	0	2	0	3	0
Gannet	69	67	0	1	0	1	0
General Dynamics	64	32	8	7	1	2	0
General Electric	12	6	2	1	0	3	0
General Motors	104	32	38	8	4	21	0

Greyhound	35	29	2	2	0	1	0
Harris	44	15	14	5	5	4	1
H.J. Heinz	33	7	14	5	3	4	1
Honeywell	82	20	34	4	8	14	1
Levi Strauss	89	57	18	9	0	5	0
Minnesota Mining and Manufacturing	33	6	17	2	2	4	2
Mutual of Omaha Insurance	5	0	0	0	0	0	0
Pfizer	36	15	8	11	3	1	1
Potlach	8	8	0	0	0	0	0
PPG Industries	11	4	2	1	1	2	1
RCA	27	17	4	1	2	3	0
Rockwell International	7	5	1	1	0	0	0
Sante Fe Southern Pacific	19	0	0	0	0	0	0
Teledyne	14	12	0	2	0	0	0
Textron	15	5	6	3	0	1	0
T.R.W.	29	6	20	1	0	1	1
U.S. Fidelity and Guaranty	0	0	0	0	0	0	0
United States Steel	42	21	7	3	10	1	0
United Telecommunications	42	40	2	0	0	0	0
Westinghouse Electric	22	18	2	2	0	0	0
Whirlpool	8	1	0	1	0	3	0
Total	1,368	684	278	96	46	127	16

see the two groups of leaders as working together as representatives of separate sectors, and would not necessarily expect corporate interest in research to explain the relationship. Instead, corporatist theory would expect universities to represent either a socioeconomic production group concerned with exploiting entrepreneurial science, or, alternatively, the interests of a strong training and preparation sector.[17] As advanced training centers, universities would be considered as vital to maintaining entrepreneurial science. For this interpretation to be viable, CEOs would have to recognize universities as partners in production or partners in personnel preparation. However, this partnership does not include elements vital to the production process, elements such as labor.

Corporate Investment in Research, Higher Education, and University Expertise

Information on corporate research and development was difficult to locate and assess. Using both *Moody's* and annual reports, I was able to locate data on research and development for 23 (74 percent) of the 31 industrial and public utility companies reported in both sources (see Table 5.3). The figures corporations presented were often in notes to the balance sheet, and indicated how research and development was treated in the accounting process. Given the tax incentives for research in recent years, corporate research and development is probably overrepresented rather than underrepresented.

The corporations' research and development effort was calculated as a percentage of net sales. In general, the corporations did not seem to devote a large financial effort to research and development. The average is 4.8 percent, and this is on the high side because, as the asterisks indicate, some of the companies making a strong effort are reporting combined company and government research and development expenses.[18] According to the National Science Board, industries actively engaged in research tend to spend from 2 to 6 percent of their net sales on research and development.[19] The corporations, then, seem not to be differentiated from other corporations in their research and development effort.

Another measure of corporate research effort is involvement in governmental research contracts. Corporations not reporting in-house research efforts did not hold large government contracts. Of the 23 corporations reporting research and development activity, 14 (61 percent)

Table 5.3. Percentage of Net Sales Devoted to Research and Development by Forum Corporations

Company	1982	1983	1984
Air Products and Chemicals	2.5	2.5	2.4
American Can	.9	.8	.9
American Telephone and Telegraph	.9	1.2	7.1
Armco	.6	.7	.7
Atlantic Richfield*	.5	.5	.6
Emerson Electric	2.6	2.7	2.8
Ford Motor*	—	—	6.1
Fuqua Industries	2.6	2.3	2.8
General Dynamics*	9.0	7.9	7.8
General Electric	2.8	3.4	3.7
Harris	14.2	15.5	16.2
Honeywell	8.6	8.6	8.1
Minnesota Mining and Manufacturing	5.3	5.4	5.7
Pfizer	5.7	6.0	6.5
PPG Industries	3.9	3.5	3.5
RCA	2.4	2.4	2.4
Rockwell International*	18.9	16.0	11.6
Teledyne	—	2.6	2.6
Textron*	6.3	10.3	10.8
TRW	2.1	2.4	2.4
United States Steel	—	.5	.3
United Telecommunications	—	.5	.3
Westinghouse Electric	5.5	5.9	6.0
Average = 4.8			

*Indicates that government-sponsored and in-house research are combined.

are among the the top 100 engaged in contract work with the Department of Defense and the Department of Energy. Twelve of the 14 are among the top 50. Although defense contracts are important to the high technology segments of many of these companies, in most cases they are not central to corporate well-being. When their "defense dependence" is calculated by determining the ratio of DOD awards to sales for 1983, only two—Rockwell and General Dynamics—are above 0.3.[20]

In sum, the in-house research effort reported by the 23 (74 percent) Business–Higher Education Forum industrials is not outstanding, but about average for American corporations. Nor is their government contract work particularly notable. Less than half (45 percent) of the corporations hold government contracts. Corporations that do not perform in-house research do not work on government contracts. Somewhat over half (61 percent) of the corporations reporting in-house research and development efforts are engaged in government contract work.

If corporations do not perform a great deal of research, perhaps they invest heavily in university research. Table 5.4 uses the *CFAE Casebook* to look at the corporations' contribution to colleges and universities, as well as to education generally. CFAE combines data for corporations and their foundations.[21] Twenty-eight (70 percent) of the 40 corporations in this sample have a gift program. The first eight columns of Table 5.4 list gifts to postsecondary education, and the last column gives total gifts to education as a whole. The postsecondary sector, which includes all categories but education, receives the largest share of gifts: $42,868,075 (68 percent). In the postsecondary category, research receives about the same emphasis (15 percent) as student aid (15 percent), and employee matching gifts receive most (28 percent).

When listing purposes they intended to achieve with their gift programs, corporations most frequently emphasized donations that would stimulate the development of knowledge and training useful to their central business. Fourteen (50 percent) of the corporate statements noted this intention. Their other major concerns were to demonstrate corporate responsibility to the community at large and to benefit the geographic regions in which they were located. They also expressed interest in benefiting the universities with which they had recruiting relationships, in aiding private colleges and universities, and in enhancing educational quality. A consistent emphasis was benefits to minorities, especially through the United Negro College Fund. Thirteen of the corporations (46 percent) voiced an intention to support minorities.

Although the corporations most frequently stated that their gifts were for research and training, their dollars were most heavily allocated to matching employee gifts to universities. This type of commitment would seem to emphasize individual alumni loyalties and to support university general funds rather than to sustain research programs.

Another way of looking at corporate investment in university research is to look at studies that examine corporate sponsorship of research. In 1980, New York University conducted a field study of companies actively supporting university research, and in 1982, the NSF performed a similar study. The National Science Board reported both studies. The New York University study showed 2 of the 40 companies in this sample as being among the top 11 performers; the NSF study showed 3.[22] In a more extensive documentation of company sponsorship of university research projects, the National Science Board showed 20 (50 percent) of the corporations in this sample as engaged in some sort of university research program. Five of the corporations were involved in five or more projects; the remaining 15, in four or less. Projects ranged from support of a professorship to funding of research institutes.[23]

Although corporations invest in university research, their investment is not great. Corporate giving programs put more dollars into matching the gifts of their employees than into research programs. The unexceptional level of corporate investment in university research confirms national trends, which indicate that in 1981 business accounted for approximately 4 percent of the dollars spent at universities for research and development.[24]

If these corporations do not engage in a great deal of in-house or governmental research, nor invest heavily in university research, perhaps they rely on academic expertise in their management decisions. To determine if this is the case, I looked at the numbers and positions of academics on the boards of directors of the corporations in this sample.[25] Of the 40 corporations, no information was available on 4, and no academics sat on the boards of 7. Thirty-six academics sat on the remaining 29 corporate boards. They represent 8 percent of the total number of directors on the boards on which they sat.

The academics were classified in one of three categories: professors, presidents, and deans or other administrators. Fourteen were professors, 12 presidents, and 10 deans and other administrators. Although the professors who sat on corporate boards represented a wide variety of fields other than the humanities, the largest contingent (35 percent) came from business schools. So too, the deans and other administrators (40 percent) were most likely to represent business or management schools. Nine of the 12 presidents were or had been active in the Business–Higher Education Forum. The university from which academics were most often drawn was Harvard, whose Business School was known as "The West Point of Capitalism," according to Philip Caldwell, himself a graduate and Forum member and president of Ford Motors.[26]

Although there is a substantial academic presence on the boards of these corporations, it does not seem significant, especially in terms of

Table 5.4. Corporate Gifts to Higher Education

Company	Unrestricted gifts	Student aid	Research	Capital	Employee matching
Aetna Life and Casualty	$ 0	$ 445,000	$ 511,000	$ 0	$ 866,068
Air Products and Chemicals	22,005	13,600	321,000	238,331	274,706
American Can	32,975	33,000	41,000	0	181,588
American Telephone and Telegraph	237,055	0	0	654,000	417,408
Armco	27,400	0	129,500	53,000	426,448
Atlantic Richfield	450,000	0	0	0	2,593,166
Carter Hawley Hale Stores	253,416	0	160,000	0	90,000
Consolidated Natural Gas	103,900	99,400	0	0	0
Emerson Electric	276,049	0	35,000	16,000	124,819
Federated Department Stores	0	0	0	541,000	254,606
First Bank System	87,244	0	0	123,382	0
Ford Motor	490,479	0	605,200	1,095,000	1,275,191
General Electric	0	1,377,300	2,499,500	0	1,619,986
General Motors	0	0	0	0	0
Harris	236,091	0	0	2,992	127,311
H. J. Heinz	0	0	0	0	324,718
Honeywell	31,350	165,052	105,732	70,000	801,662
Minnesota Mining and Manufacturing	675,650	3,685,000	120,000	180,000	289,173
Pfizer	82,000	0	411,050	70,000	315,557
PPG Industries	210,000	30,000	0	0	219,417
RCA	39,700	263,144	401,700	110,000	17,000
Rockwell International	769,000	0	0	0	335,000
Sante Fe Southern Pacific	108,450	0	0	21,000	199,562
TRW	178,956	74,053	960,716	87,000	716,336
United States Steel	0	150,000	0	0	0
United Telecommunications	3,500	43,000	35,000	0	14,515
Westinghouse Electric	0	0	0	0	526,907
Whirlpool	59,420	8,400	12,000	0	123,360
Total	$4,375,540	$6,386,949	$6,348,398	$3,261,705	$12,134,554

				Rank*
Aetna Life and Casualty	$ 0	$ 0	$ 0	$ 1,839,848	F-2
Air Products and Chemicals	0	0	0	1,023,282	I-212
American Can	11,000	47,700	45,251	593,334	I-117
American Telephone and Telegraph	0	142,200	0	2,238,025	
Armco	0	89,437	0	1,138,520	I-87
Atlantic Richfield	0	0	0	11,752,211	I-12
Carter Hawley Hale Stores	0	0	0	503,416	R-24
Consolidated Natural Gas	0	50,150	0	0	
Emerson Electric	0	69,350	0	1,298,145	I-111
Federated Department Stores	0	15,000	0	974,560	R-7
First Bank System	0	0	0	423,356	B-17
Ford Motor	266,700	260,400	0	4,273,486	I-4
General Electric	0	0	195,000	11,182,094	I-10
General Motors	0	0	0	5,152,000	I-2
Harris	0	12,850	0	389,725	
H. J. Heinz	0	0	0	1,029,078	I-100
Honeywell	0	147,850	1,104,000	2,824,517	I-60
Minnesota Mining and Manufacturing	0	277,750	3,903,370	2,234,574	I-47
Pfizer	0	0	0	1,305,378	I-99
PPG Industries	0	30,000	0	1,187,594	I-103
RCA	0	0	68,800	2,050,000	S-2
Rockwell International	0	130,000	0	2,200,000	I-43
Sante Fe Southern Pacific	0	207,000	0	728,539	T-1
TRW	0	118,290	241,950	2,819,786	I-63
United States Steel	0	122,500	1,325,000	237,748	I-215
United Telecommunications	8,000	0	0	148,614	U-30
Westinghouse Electric	0	0	1,283,600	2,330,507	I-34
Whirlpool	0	170,775	17,006	825,556	I-147
Total	$285,700	$1,891,252	$8,183,977	$62,703,893	

*The rank is the corporation's position in the *Fortune 500*s. "I" is the industrial 500, "R" is the retail 500, "B" is banks, "S" is service, "U" is utilities, "F" is financial, and "T" is transportation.

research. If these academics were valued for their scientific ability, one would expect them to be drawn from fields such as biotechnology, physics, computer-related areas, even engineering. None of them represented these specialties. These academics may embody training rather than research, linking, as they do, preparation with practice, as members of business schools and policy forums.

The presidents are not listed by specialty, and probably do not actively link either research or training into the deliberations of corporation boards. However, many of the presidents sit on a number of boards other than those in this sample.[27] Through their directorships, the presidents may broadly represent the university community to the business world.

Research does not seem to provide an obvious basis for cooperation or co-optation in corporate and university relations. The corporations in the Forum do not appear to be heavily engaged in in-house high technology research. Nor do they invest heavily in university research, nor make extensive use of academic expertise. The data, then, confirms neither a functionalist research-for-resources interpretation nor a neo-Marxian interpretation in which universities' high technology research is co-opted for corporate use. Both interpretations depend on corporations displaying a greater interest in advanced research than the data suggest.

Social Responsibility

If neither research investment, nor commitment to higher education as manifested through giving, nor heavy use of expertise serves as the basis for corporate cooperation in the Forum, perhaps the CEOs are a socially active segment of corporations, committed to broad progressive goals, from which support for higher education could be deduced. A hallmark of professional managers is that they are supposed to keep the common as well as the corporate good in mind.[28] I used the *Concerned Investors Guide* to see how responsible the corporations in the sample were in discharging their obligations to society as a whole.[29] The *Guide* rates corporate performance in the following categories: environment, fair labor, occupational safety and health, product safety, antitrust, civilian nuclear industry, and activity in South Africa.

I calculated the number of allegations of regulatory violations as well as the number of convictions made. The years for which information was analyzed were 1978, the year the Forum was organized, to 1983, when the *Guide* was published. When it was not possible to discover the disposition of complaints—as was the case with the Environmental Protection Agency—I did not use the category. I understand, of course, that the outcome of a complaint might say as much about the competence of the plaintiff's attorneys or the administrative philosophy of the presiding judge as about the performance of the corporation. With that caveat, I am nonetheless going to use pleas of no contest on the part of the company, convictions, consent decrees, or decisions against them as an indication of company neglect of societal well-being.

Companies were most frequently involved in complaints made to the Occupational Health and Safety Administration. Of the 35 companies in this sample indexed in the *Concerned Investors Guide*, 24 (68 percent) had complaints made against them. Altogether 188 complaints were lodged, of which 57 (30 percent) were uncontested.

Nineteen (54.2 percent) of the companies were involved in 82 fair labor practice cases that were listed in Bureau of National Affairs' publications, *Fair Employment Practices* and *Labor Relations Manual*. Twenty-two (26 percent) of these cases were decided against the companies.[30] The largest number of cases centered on violations of the Civil Rights Acts, with regard to racial discrimination. Thirty-eight complaints were made, and nine (22 percent) were decided against the company. Other complaints involved discrimination in terms of sex, age, national origin, and handicap, as well as complaints of reverse discrimination.

Fourteen (40 percent) of the corporations were involved in cases where product safety violations were brought before the Consumer Product Safety Commission and the Food and Drug Administration. There were a total of 55 cases, of which 52 (94.5 percent) called for remedial action on the part of the company—for the problem with the product to be corrected, or for the product to be recalled, modified, or replaced. The most common remedy was recall, which occurred in 25 (50 percent) of the cases.

Twelve companies (34.2 percent) were included in antitrust suits reported by the Justice Department. These twelve were engaged in 22 cases. Eleven (50 percent) agreed to consent decrees, in which they will abide by the decision or settlement of the court. Five (22.7 percent) were decided in the companies' favor. Six (27.2 percent) found the companies guilty of price fixing, and fines were levied.

Fourteen (40 percent) of the companies accounted for a total of 18 operations in South Africa. Using the report prepared by Arthur D. Little, Inc., for the Reverend Leon H. Sullivan of the International Council for Equality of Opportunity Principles, the *Guide* offered the following information for 1982: 2 of the companies were not signatories to the Sullivan Principles; 1 was a signatory that did not report; 1 was a signatory that did not pass the basic requirements for Sullivan Principles signatories; 1 recieved a low point rating; 2 were endorsers of the Principles, but had no employees. By far the largest group was listed as making progress (7) or as making good progress (4).

The corporations' social responsibility record does not readily lend itself to interpretation. In the area of occupational health and safety, the instances in which corporations did not contest complaints are probably admissions of guilt, but this occurred in only 30 percent of the cases. Corporations may have pleaded no contest only in situations where gross negligence was involved, and fought regulation when minor infractions were present. What is clear is that occupational health and safety is an arena in which employers and employees are struggling to define what hazardous working conditions are.

Approximately half of the companies were involved in fair labor practice cases, with roughly a quarter being cases of racial discrimination, of which 22 percent were decided against the company. Given the companies' interest in gift programs that benefit minorities, discussed earlier, this is somewhat surprising. The companies seem to be engaged in contradictory practice, on the one hand supporting minorities through educational investment, and on the other discriminating against them in terms of work.

Although only 40 percent of the corporations were involved in product safety cases, the number in which remedial action was involved is alarmingly high. Certainly poor worker performance accounts in part for product safety violations, but cases brought against the companies suggest at the least a lack of interest in close inspection on managements' part. Whoever is to blame, the situation seems to give credence to the CEOs concern, discussed later in this chapter, that low-quality goods are blunting our competitive edge in global markets.

Although the involvement of corporations in antitrust cases is not particularly high, the instances in which they are at fault or partially so are significant. The relatively low number of cases may reflect societal acceptance of oligopoly, if not monopoly. Conversely, the low number of cases could suggest a lack of violations. That reading of the data, however, is contradicted by the high number of cases in which corporations are at fault. The instances of fault, together with the low number of cases, may reflect the current climate of deregulation, in which rules are not clear.

Overall, the record does not indicate that the Forum corporations are characterized by an umblemished record of social responsibility. Perhaps the only thing that can be said with certainty about the record of this sample is that its concrete response to regulatory tribunals has been to contest their authority. In this, they are like the great majority of large corporations.

Concerns of Corporate Leaders

I thought that an analysis of articles, speeches, and commentaries written by CEOs who are heads of Forum corporations might provide a direct indication of why they are interested in participating in organizations like the Forum. As several scholars have noted, corporate leaders began bringing their views more aggressively to the public in the early 1980s, in an attempt to shift the national climate of opinion to more fully support business.[31] I hoped that in making their case to the public they might address their interest in high technology, higher education, science, and research.

The 46 corporations treated in this chapter had 58 CEOs active in the Business–Higher Education Forum between 1983 and 1987. Using their names, I did a computer search of the Magazine Index for the years 1978–1986, the years in which the Business–Higher Education Forum was in operation. The Magazine Index covers general periodicals in the areas of news, business, and the several sciences as well as trade publications. Twenty-two of the fifty-eight CEOs (40 percent) wrote articles in magazines covered by the Index. All told, they were responsible for 62 articles in 34 publications.[32] They wrote most frequently for general business and management magazines. These publications, such as *Design News*, *Industry Week*, and *Fortune*, accounted for approximately one-third of their articles. The CEOs were also regular contributors to trade magazines, such as *American Metal Market*, *Public Utilities Fortnightly*, and *National Underwriters*. About 30 percent of their articles appeared in trade magazines. They also wrote letters to and columns in newspapers (13 percent), articles for news magazines (6 percent), and had their talks appear in *Vital Speeches* (6 percent).[33] One even wrote an article for a policy journal, *Foreign Affairs*.

The material they published was generally short. About half appeared in opinion pages, running from a single column to two pages. Articles between three and five pages were often versions of material that had appeared elsewhere, such as speeches or public remarks. Only one article was more than five pages. The number of contributions to the public press made by CEOs increased over the years under consideration. Forty-four articles (71 percent) were published between 1983 and the present.

In their publications, the CEOs usually addressed economic problems and proposed broad solutions to them. Competitiveness in the global market was the topic most often addressed, with 27 articles written by nine CEOs. How to create the successful corporation of the future was next, with 12 articles by eight CEOs. A variant of the successful corporation of the future theme, the "success story," gives accounts of reform that led to corporate triumphs in the global market. There were 4 such articles, by four CEOs. The problems of specific industries were addressed in 9 articles, by five CEOs. Science, research, and development were the topic of 3 articles by two CEOs. One article addressed higher education. Six addressed widely disparate themes, and were categorized as "other."[34]

The problem of competitiveness in the global market was paramount. There was general agreement on the definition of the problem: the United States was no longer the premiere world economic power. The problem was seen as stemming from both domestic and international difficulties. In the domestic arena, U.S. manufacturing suffered from falling productivity and uncertain quality, high labor costs and high inflation, and the budget deficit. In the international arena, the United States faced increased competition, especially from newly industrialized nations, and unfair trade practices—such as dumping abroad, protection at home, disregard of intellectual property laws—by these same nations. Compounding these difficulties were the trade deficit and the international currency market, which kept the dollar high, especially in relation to the yen.

There was a high level of agreement about how to hone America's competitive edge in the global market. The world market had to be smoothed into "a level playing field" in which no nation would have unfair advantages.[35] In other words, CEOs were advocating the creation of a global free market in which the United States would have to submit to the same market discipline as any other country. If we were unable to compete, our industries would get no special help. Although the executives seemed to think that the United States would be able to maintain its competitive edge in the international arena, they perceived, in Ruben F. Mettler's (TRW), words, "the world market... [as] a harsh task master."[36]

Ironically, a level playing field for international free trade could not be created without sustained government intervention. The CEOs did not have uniform prescriptions for specific government interventions, but were agreed that critical to competitiveness was "a more cooperative business-government relationship. Our goal should be to eliminate the adversary element in our dealings with each other."[37] Their particular proposals for government action covered a wide spectrum: the creation of incentives for aggressive corporate action in the world market, including policies aimed at increasing savings for investment in business; implementation of policies that would contribute to capital formation in the United States; adjustment of international currency ratios; reduction of disparities between national economic systems, including tax systems; and greater investment in research and development. Although there was not a great deal of consensus or conflict around any one of the above policies, the CEOs did share a commitment to two policies. First, most of the CEOs believed that government "overregulation" had to be curbed in the areas of consumer and worker protection and in the achievement of social goals.[38] Second, they were convinced that domestic antitrust law enforcement had to be reconciled with the reality of the world market. Such reconciliation would permit large corporations to pool their assets—in research and development as well as in capital—to better position themselves in world markets.

Although the government had the responsibility of leveling the playing field of the world market, CEOs were also willing to do their part to increase competitiveness. By and large, CEOs were agreed that manufacturers had to improve quality, perhaps by more intensive use of "quality circles," that the traditional adversarial relationship between labor and capital had to be changed, and that top-heavy management structures had to be reformed. Accelerated automation was also necessary for successful competition.

The CEOs believed that the American people had to come together in the spirit of cooperation and play a part in the competitive struggle. In the words of Philip Caldwell (Ford), "A critical first step would be more widespread recognition by the American people that we do have a real problem."[39] The creation of consensus among the central sectors of American society was critical. As Robert Anderson (Rockwell International) said, "I am confident that if management, labor, government and academia all work together, America will regain its position as the world leader in productivity, quality, and trade."[40] Philip Caldwell (Ford) reiterated this theme: "We must have a strong, balanced economy—in real, not inflationary, terms—if we are to be effective in the world. That isn't just a business interest, but a major national interest."[41]

When CEOs spoke about the future of the corporation, they touched on many of the same themes as they did when addressing competitiveness in the global market. To reach the future, the obstacles of the moment had to be overcome. America had to keep a competitive edge, surmount decline, raise falling productivity. Many of the solutions too were reiterated. Capital and labor had to end hostilities, quality had to be built into the production process, more money needed to be spent on research and development, government regulation had to be curtailed, and a global free market had to be established and maintained.

However, the futures that captured CEOs imagination were not all the same. Certainly most shared a vision that looked to high technology, coupled with corporate reform. In the words of Edward Donley (Air Products and Chemicals), high technology would enable us "to provide...leadership for ourselves and our allies well into the 21st Century."[42] But William S. Woodside (American Can) saw the possibility of "a 100% service economy...aside from defense."[43] And David M. Roderick (U.S. Steel) argued that high technology should complement rather than replace basic industry, although he conceded that the United States would probably never again dominate the global marketplace in basic industry.[44]

Five of the eight CEOs who spoke of corporate prospects talked about the "factory of the future." The factory of the future was almost wholly automated, peopled with robots rather than workers. When Roger B. Smith (GM) spoke about "get[ing] on-line with the future," he envisioned computers bearing the brunt of the work in design and engineering, of "computers in the offices and factories [that] will finally be able to 'talk' to each other." And "on the factory floor, the network of automated processes will be computer controlled and checked by machine vision technology. Some robots will be programmed to respond to voice commands." The dealership sales people and service technicians "will be trained and updated through the corporation's telecommunications network. Service will be simplified through such diagnostic tools as artificial intelligence."[45] The factory of the future was the way to maintain our competitive edge in global markets.

The factory of the future called for corporate reform. Many of the same themes were sounded as in their discussions of competitiveness: a commitment to quality, to increased productivity, to new relationships between labor and management. CEOs spoke of forming management practices as well as labor, and emphasized "streamlining." William S. Woodside proudly announced that he had already reduced the managerial staff of American Can from 1,200 to 250, and wondered if he could go to 150.[46] Senior managers would be asked to participate in product creation, using the new technology. Another concept discussed was "leap-frogging" to the factory of the future. Leap-frogging meant using any technology available to move to the generation beyond the one on which our competitors were working. Diversification was viewed as important, especially in terms of high technology fields. Joint ventures with companies in other nations were presented as another way to conquer world markets.

These strategies for reform were essentially amplifications of the practices in which successful corporations were currently engaged. The four CEOs who told their "success stories" spoke about similar ideas of corporate reform. They talked about reorganization, diversification, selective acquisition, especially in high-technology fields, and joint ventures. In essence, they talked of a systems approach to the world market, in which multinational corporations could design, source, manufacture, and sell wherever the price was right.

CEOs were not totally focused on questions about competitiveness in the global market and in the future. A number addressed specific problems faced by their industries. The industries were insurance, health care, chemicals, steel, electricity, and telecommunications. In the main, these articles addressed the role that consumer, state, and corporation would play in bearing industry costs. In service sectors, such as insurance and health care, the consumer was expected to bear a greater share of the costs. In older industries, such as chemicals and steel, the government was expected to pick up a larger part of the tab. In utilities, the position on government intervention varied according to the degree of deregulation. In all cases, corporations were concerned with moving costs away from the private sector.

Education did not seem to be a major focus of CEOs as a means of improving American competitiveness, whether in terms of enhancing managerial performance or employee skills, building industry-university partnerships, or creating the technologies of the future. Of the 62 articles, 3 dealt explicitly with science, and 1 with higher education.

Another 9 articles, written by seven CEOs, mentioned these subjects, with 3 discussing them at some length.

Corporate leaders realized that they shared common interests with university presidents on the subject of regulation. As Gerald Laubach (Pfizer) said:

> There are demonstrable, important direct links between social attitudes and science policy. Agitation and alarm about science and technology have tended, almost without exception, to spawn political reaction. That, in turn, tends inevitably toward new or more regulation—in other words, toward increased *political* control and direction of the scientific enterprise [emphasis in the original]... Even our universities—bastions of academic freedom—have not been immune to the impositions of politically motivated controls on their research activities... Indeed, all scientists need to become more alert to the reality and to the danger of advocacy science, or politicized science, or to the science of confrontation... By their very nature regulations set *a priori* standards for scientists to adhere to.[47]

Laubach, president of a pharmaceutical company, saw regulation of science as creating problems for corporations as well as for the university, and noted how regulation had limited the sale in the United States of a host of life-saving drugs used in Europe, and threatened "Draconian controls" on recombinant DNA research. He was deeply disturbed by government infringement of managerial prerogatives, whether in science or industry.

Admiral "Bobby" Inman, who became a member of the Business-Higher Education Forum after he left the CIA to be CEO of the Microelectronics and Computer Technology Corporation, wrote two articles in the early 1980s that addressed government regulation of research related to national security issues. Rather than government review and classification of research after completion, Inman argued for incorporating the question of potential harm to the nation into the peer review process, prior to the start of research and publication.[48] This was regarded as a moderate position, given that a faction in the executive branch was considering amending the Arms Export Control Act to enable the government to impose controls over all technical information, whether in the persons of scientists attending international conferences at which there was a Soviet bloc presence, or in the export of high technology equipment which might reach the Soviet bloc. Although Inman was concerned with limiting government regulatory powers in corporate or academic science, his position was not as protective of academic freedom as was that of university presidents. As we saw in the previous chapter, university presidents wanted clear rules rather than self-censorship.

Inman's position on antitrust regulation was similar to that taken by university presidents. When he became head of the Microelectronics and Computer Technology Corporation (MCTC), he wanted to create a research consortia of electronics firms interested in developing technologies that would enable them to beat the Japanese in the race to develop a fifth-generation computer. MCTC was ruled "precompetitive" and therefore allowable.[49] University presidents shared Inman's interest in the creation of research consortia because these organizations were potential funders of university research.

Business executives were concerned with forestalling government regulation that they believed infringed on managerial decision making and trade freedoms. They saw themselves as sharing a common ground with academics on this issue, and, in the name of autonomy, urged both communities to resist further external encroachment. However, not many CEOs addressed this issue.

The three CEOs who mentioned science, research and development, and education in more than a passing fashion all dealt with the critical importance of research and development for American competitiveness. Edward Donley (Air Products and Chemicals) looked to the history of industry and government cooperation in research and development. He noted that government had traditionally engaged in sponsorship of research critical to the national interest when immediate economic rewards did not justify the financial risk for private sector investment. He recommended continuation of this pattern of funding.[50] Edson W. Spencer (Honeywell) suggested a more comprehensive approach to research and development, one in which the federal government considered setting long-range goals and strategies for the nation's future, targeting "R&D for all agencies to accomplish these objectives in broad terms, rather than negotiating R&D on an annual basis through countless line items in the budget."[51]

A single article dealt with the kind of higher education in which CEOs were interested. In "What's Good for General Motors: Liberal Arts," Roger B. Smith (GM) argued that a liberal arts education best served high-level managers, such as those who ran the *Fortune* 500 companies. He valued the conceptual abilities, creativity, and commitment that came with a liberal education. These skills, he argued, were invaluable to corporate leaders of the future, who would have to operate in an environment of rapid and constant change and competitiveness.[52] Smith did not want the technical, research, or professionals skills and competencies often emphasized by those promoting university-industry-partnerships, but looked instead to a traditional curriculum to provide the breadth of learning and depth of conceptual thought necessary for focusing the energies of the whole person on the goals of the competitive corporation.

The CEOs articles in the public press emphasized topics, touched upon Chapter 2, that initially concerned the Business-Higher Education Forum, and prefigure a number of issues to which the Forum later turns, as shown in the following chapters. Competitiveness in global markets is the overarching and organizing theme throughout the CEOs' material in the public press and is a persistent concern in the Forum. Another theme that informs the CEOs' comments in the public press and that pervades Forum documents is the importance of using the state to promote business interests. The state must intervene to set the rules of international competition; it must coordinate private sector interests more effectively domestically; it must work more cooperatively with business. The CEOs and Forum leaders expect to give direction to state infrastructure created to meet private sector needs. Another theme that spans the CEOs' comments in the public press as well as later Forum material is the importance of changing anti trust regulations to enhance corporate ability to compete abroad. Yet another theme is the need to do away with cross-industry regulation to unfetter corporate energies. These themes—competitiveness, using the state to organize business interests, reshaping anti trust legislation, and doing away with cross-industry regulation—are emphasized and re-emphasized in the Forum.

A number of phrases used in the CEOs' material for the public press appear in the Forum documents, where they are expanded, sometimes as major themes, other times as motifs. Among these are a concern with representatives of "management, government, labor and academe" working together to resolve the economic crisis caused by global competition, and a firm resolve to end "adversarial relations between labor and management." Taken together, these phrases suggest a stress on consensus building, thought to be central to maintaining a competitive edge.

Other phrases used in the Forum that echo the CEOs' voice in public press are an interest in rapid development of "the factory of the future," and a perceived need on the part of business leaders to reorganize so there is "no top heavy management." In the following chapters, through the Forum documents, the meanings of these phrases are more fully explored.

The CEOs' comments in the public press give no indication of shared social activities or ties outside their business interests, broadly construed. Nor, as noted in Chapter 3, is there evidence of ties on the part of these CEOs to an upper class defined by wealth or ownership of the means of production. However, the positions taken by Forum CEOs in the public press do suggest, in Useem's words, that they are men "who share with other corporate managers a common commitment to enhancing company profits [but whose] heightened sensitivity to business interests more general than those that look solely to support individual company profits also sets them apart."[53] They seem, then, to have a classwide consciousness of corporate interests.

University presidents' testimony before Congress sometimes stressed the same materials CEOs presented in the public press. In the preceding chapter, we saw university presidents speaking to the importance of increasing global competitiveness, reducing cross industry regulation, mitigating the effects of anti trust legislation in the domestic arena and building the factory of the future. But CEOs' testimony rarely touched the interests of university presidents. They did not speak strongly to enhancing corporate research or providing monies for university research, whether basic or applied, nor to building business-higher education partnerships. When they turned their attention to these matters, they most often looked at ways in which research and development could benefit business. CEOs did not mention increasing financial support for higher education generally, nor maintaining student financial aid. In short, they did not often speak to the topics or themes that preoccupied presidents. Presidents, as indicated above, more often spoke to topics of interest to business.

Conclusion

On the one hand, the public remarks of CEOs do not confirm the pluralist resources-for-research model as a basis for cooperation, nor the neo-Marxist research-as-point-of-co-optation model because the CEOs by and large do not speak to these issues. On the other hand, the CEOs' remarks do not refute these models, since it is possible that they deal with the basis of cooperation or co-optation elsewhere. However, the CEOs' lack of attention to the university and its concerns does not suggest the vital relationship that participation in the Forum implies.

There is a possibility that research is not the critical link between corporations and universities, at least as far as corporate leaders are concerned. As the preceding chapter indicated, presidents of research universities strongly stress research as a link between business and higher education. Their emphasis on research may reflect their concerns with developing entrepreneurial science, perhaps to the point where they have a good deal of independence from both corporations and the state. CEOs may not be supportive of this agenda, and could even see it as competitive. They may value their relationship with the university because of what the academy does for corporations in terms of training and preparation across a wide spectrum of professional careers, and for the development of personnel for the maintenance of established science-based industry.

If CEOs work with university presidents to maintain and enhance more traditional relationships, functionalist and neo-Marxian analyses would take a slightly different cast. Functionalists would probably explain the relationship in terms of an exchange of training and preparation for resources, with the professional goals embodied in career preparation having as much weight as corporate considerations of profit. Neo-Marxists would probably see professional goals subordinated to but legitimating the hegemony of the dominant class. Neither of these analyses would treat university presidents' pursuit of entrepreneurial science, as embodied in their congressional testimony, and corporate leaders' seeming lack of response to it.

Corporatist theory can accomodate links between business and higher education regardless of whether they are made through training and preparation or through entrepreneurial science. Rather than seeing the ties between the sectors in terms of functional relationships of equal exchange, corporatist theory would look for a conscious building of partnership. The partnership would be centered on production, whether the university engaged in preparing students to work in the business sector or whether the university itself engaged in production. Unlike neo-Marxist theory, corporatist theory would not see university interests as subordinate to those of the business sector, although corporatists would expect the needs of production to be dominant. Corporatist theory fits well with the partnership business and higher education are forging in organizations like the Forum, but other groups, such as labor and agriculture, theorized by corporatists as critical to production, are generally not referred to in the speeches of university presidents, the remarks of CEOs in the public press, or the Forum documents.

CHAPTER 6

The Business–Higher Education Forum Reports: Corporate and Campus Cooperation

In this chapter and the next, I look at the reports produced jointly by corporate and university leaders in the Business–Higher Education Forum, focusing on what they say about the Forum members as political actors.[1] After reviewing the reports, I try to see which of the several theoretical traditions—pluralism, Marxism, corporatism—best explains the degree of cooperation or shared consciousness between the two sectors, the policy positions the leaders develop, their expectations of and attitudes toward the state, and the ideology they promulgate. At this point, I will briefly review the position the several theoretical traditions take on the role of the state and ideology in policy formation so that these questions will be in mind when the Forum documents are analyzed.

Neopluralists see the state as quite distinct from civil society, although they recognize a state technocratic substructure. The state is viewed as partially controlled by electoral politics, interest group pressures, and media attention, and partially by economic pressure mounted by business and usually directed toward the executive branch. However, the power of business is thought to be confined to matters that touch directly on the maintenance of capitalism and its prerogatives. In terms of most other issues, polyarchy prevails and looks quite like conventional pluralism. There is no inevitable ordering of power; business does not necessarily dominate in areas not directly related to its immediate interests. Ideology is not seen as central to the various interest groups trying to influence state power; indeed, ideology very often interferes with the rational development of state policy.[2]

Neo-Marxist theories of the state take a quite different position. Rather than seeing the states as separate from civil society, they see the state as "an expression, or condensation, of social-class relations, and these relations imply domination of one group by another. Hence the State is both a product of relations of domination and their shaper."[3] The state has a degree of autonomy, or relative autonomy, which is essential for legitimacy and popular support, but this autonomy creates contradictions, "bringing the class struggle into the political apparatuses," and creating the "possibility of subordinate classes and groups taking over these apparatuses, thereby interfering with the class-reproductive functions of the State."[4] Ideology is central to the class struggle model, because state apparatuses generate and disseminate ideologies which make credible the notion of an autonomous rather than a class state, thereby sustaining the class state.

Corporatist theory sees the state as "a constitutive element engaged in defining, distorting, encouraging, regulating, licensing and/or repressing the activities of associations—and backed in its efforts, at least potentially, by coercive action and claims to legitimacy."[5] State relations are patterned not by class interest but by interest associations, or sectors, which minimally include those engaged in the productive process. Relations between the sectors are defined not by conflict, but cooperation. The sectors endeavor to work harmoniously to create policy and sometimes even assume state responsibility for policy formation and implementation. Corporatist theory does not sharply distinguish the state from civil society.

Methodological Considerations

With the several theories serving as a frame for discussion, I will turn to the Business–Higher Education Forum reports. The reports are documents growing out of the deliberations between CEOs and presidents over shared problems. In examining these documents, I tried to establish the ground on which corporations and universities are building common interests. Understanding the degree of shared consciousness between the groups is critical to assessing their sense of themselves as policy actors transcending their separate sectors, to evaluating the policy positions they develop, and to understanding their posture toward the state. Although there is a long history of cooperation between corporation and campus, there is also a substantial history of opposition to cooperation, or of deliberate distance.[6] Because cooperation between the two sectors cannot be taken for granted, the points selected for maintenance of the relationship become theoretically interesting. Does cooperation center around research and and development partnerships, a dedication to high technology and its applications, the routine demands of science-based industry, training, and legitimation, or points to which not much attention has yet been paid?[7]

In looking at the Forum's documents, I attempted to develop content analysis techniques that focused on points of convergence between corporation and campus. For each report, I looked at the ratio of university concerns compared with corporate as expressed in the text, ideological structure (structure of the argument, major themes, rhetorical structure), and the disposition of societal resources as indicated by recommendations.

The ratios were established by calculating the inches of text devoted to the concerns of each sector.[8] An examination of the ratio of university concerns as compared with corporate gives a sense of the degree to which the Business–Higher Education Forum is a partnership. The ratios also provide a rough measure for determining the areas in which the interests of one or the other participant dominates.

Ideology was explored by looking at several structures running through the documents: the structure of the argument, the major themes, and the rhetorical structure. The structure of the argument was taken to be the logical line of reasoning that moved from problem definition to conclusions.[9] An analysis of the structure of the argument allows an evaluation of the rationality and cohesiveness of the program of action advocated. Themes emerged from a substantive content analysis and encompassed the prescriptive amalgam of narrative, statistics, and indirect recommendations that make up the body of the reports.[10] The major themes depict content and reoccurring concerns. The rhetorical structure, built by figures of speech as well as by feeling and value-charged language, allows a glimpse of the emotional argument being made and the central values held.[11] Taken together, the structure of the argument, the themes, and the rhetorical structure reveal the ideological underpinnings of the report. If ideology is taken to be the ideas, values, and beliefs that order our sense of who should get what and what moves us to action, then the Forum policy documents should be useful in determining commonalities between business and higher education.[12]

The disposition of societal resources was determined by looking at whom the specific, action-oriented recommendations or initiatives advanced in the reports are likely to benefit.[13] Looking at how the resources are allocated allows a rough evaluation of the ideological claims made. It is also possible to see if priorities are given more than lip service.

As a way of approaching an organization such as the Business–Higher Education Forum, a systematic examination of its official documents certainly has limitations. This method misses the politics that go into producing the document, the intensity of participation, the influences of authorship, and covert agendas of all sorts. However, a careful content analysis sensitive to the logical, empirical, emotional, and ideological layers of the text does let us see the formal policy positions to which CEOs and university presidents assent and gives us glimpse of the direction they want to go and a sense of which sectors of society will reap the benefits and which will bear the costs. Although the men and women who participate in the Forum may not have shared in the actual writing of the policy documents produced, they are knowledgeable, sophisticated leaders in their respective spheres, careful of their reputations, reflective about how they use their influence, and well aware of what an endorsement means.

In this chapter, the content analysis techniques described are applied to the first two Forum reports, *America's Competitive Challenge: The Need for a National Response* and *Corporate and Campus Cooperation: An Action Agenda*. These two reports are considered in great depth, the first because it outlines the basic philosophy and goals of the Forum, and the second because it lays out the relationship between corporate and university communities, describing conceptions of reciprocity, together with mutual rights, duties, and obligations. In the following chapter, the rest of the Forum reports, four in all, are analyzed. At the end of the next chapter, the Forum reports in toto are considered in light of the theories of the state to see which best explains the Forum's efforts in the policy process.

America's Competitive Challenge

In 1982, President Reagan invited the Forum "to prepare a set of recommendations designed to strengthen the ability of this nation to compete more effectively in the world marketplace."[14] The report, *America's Competitive Challenge*, was the Forum's response, formulated by "representatives of America's business and academic communities." Although developed conjointly, the report's major focus is not on the higher learning. All told, the report was a total of 191.75 text inches, of which

34.25 were devoted to university activities, research, education, and training across the several levels of American schooling. The ratio of university and educational concerns to corporate expressed in the text is 1:6. The concerns of the university, then, do not dominate the report.

The structure of the argument in *America's Competitive Challenge* is aimed at answering the question: What policies can be developed "to improve the ability of American industry and American workers to compete on an international scale"? According to the Forum, the United States is falling behind on most worldwide measures of productivity and market shares. The American economy is now "inextricably linked to the international economy," and unless the United States can compete successfully in global markets, our economy will not be revitalized.

The report rests on a single key assumption: the private sector is the engine of progress, and only by acknowledging its primacy will recovery be possible. "*Thus, the Forum makes only one overall recommendation: as a nation, we must develop a consensus that industrial competitiveness is crucial to our social and economic well-being. Such a consensus will require a shift in public perceptions about the nature of our economic malaise.* (italics in the original)."

The premises, or assertions, are several. U.S. problems are deep-rooted and structural: no "quick fix" is possible. The text indicates that quick fixes are solutions that would diminish the power of business leaders to act in the global market place. One quick fix is protectionism, the other is centralized direction of the economy. Neither is acceptable.

To revitalize the American economy, a consensus must be reached among divergent groups on the importance of remaining globally competitive. Private sector groups—industry, higher education, and labor—as well as government and the public must work to build consensus. The first item on the cooperative agenda for these groups is recognizing the primacy of business and moving forward to respond to the economic challenge raised by other countries.

High technology or "new technologies" are viewed as the key to successful American competition. High technology increases labor productivity in a way that is impossible in older industries. The United States has a slight advantage in terms of high technology, given the nation's strong scientific base.

Although each sector—business, higher education, labor, and government—has its prescribed role in revitalization, the role of federal government, in the person of the President, is most exactly specified.

The President should play a central role in fostering U.S. competition in global markets, beginning with a public address that emphasizes increased competitive capacity as the only real solution to economic problems. He should appoint an advisor on economic competitiveness, with cabinet-level status, and should establish a National Commission on Industrial Competitiveness. The Forum also recommends that the U.S. Department of Commerce establish a Bureau of International Competition.

All in all, the structure of the argument emphasizes creating a framework of ideas and institutions, dignified by presidential approval, that stresses dedication to business competition in world markets and provides a wide range of support—symbolic, political, material, fiscal—to multinational corporations. The major themes running through the report are five: the primacy of business, deregulation, redistribution, privatization, and the centralization of policy making.

The primacy of business, as noted earlier, is an assumption as well as a theme, the point on which the logic of the report turns. Robert Anderson, CEO, Rockwell International, and David Saxon, president, University of California, put it thus in their letter of transmittal to Ronald Reagan, President, United States: "And unless we rebuild the American economy and strengthen our educational system, it will be increasingly difficult—if not impossible—to maintain a just society, a high standard of living for all Americans, and a strong national defense."

Deregulation is a persistent theme, with special attention given to those agencies responsible for cross-industry regulation. Specifically, regulatory agencies are asked to become less antibusiness; environmental and land use regulations should be loosened in the utilities and petrochemical industries; "technology forcing" regulations under the jurisdiction of the FDA and EPA should be modified so that the "delicate process of innovation" is not "overwhelm[ed]." Antitrust laws are also singled out for attention, even though on the surface it appears that such legislation fosters the competitive posture the report champions. However, antitrust laws, like regulatory practice, constrain the freedom of business leaders to make decisions to strengthen multinational corporations operating in the global arena and interfere with business leaders' opportunity to pool resources, especially with regard to research.

Redistribution involves shifting public resources from social welfare programs to business welfare. This transfer of monies is not presented as zero-sum game in the report, but the amount of funds for business programming, given the deficit and the state of the economy, indicates a marked shift in priorities. Substantial government monies are requested for corporate as well as university research, for changes in tax laws to encourage modernization of plants and equipment, and for broad education and retraining programs.

Privatization calls for moving to the private sector what was once part of the public domain. Privatization centers on research and intellectual property. The report asks that practices established in the 1940s, when government first engaged in the heavy funding of research, be changed to allow discoveries made through federally funded science to be patented privately and exploited by universities or businesses. It also calls for intellectual property laws at home and abroad to be tightened so that more and more material—ideas, processes, formulas—is treated as private property.

Another theme is the centralization and concentration of government powers influencing business policy. The development of a foreign economic policy is called for, as is streamlining of existing agencies. Streamlining involves the development of a regulatory system that integrates federal, state, and local decision making. Also advocated is the creation of a single agency to deal with human resource problems stemming from dislocation due to changes in the economic infrastructure.

Generally, the themes stress the primacy of business. Cross-industry regulation constraining the decision-making power of multinationally competitive business should be modified. Antitrust laws should be altered to allow for greater monopoly at home to better enable corporations to compete abroad. Knowledge once part of the public domain needs to be commodified to allow for more private ownership of ideas and the technology that stems from them. Public resources have to be redistributed to give business a greater share. Government decision making about business has to be centralized and streamlined. Although the report begs the question of who will head this decision-making process, the framework established gives a strong indication of the answer. The themes, then, deal with a change in priorities and a transfer of resources that enable the heads of large business to oversee the shifting of the American economy from a domestic to a global arena.

The rhetorical structure of the report is informed by three central metaphors—competition, consensus and "disordered public policy." The metaphor of competition pervades the report. Competition is what we face in the global market. Rather than the rule-governed, good-spirited competition of the playing field, the competition invoked is the "cutthroat" competition between businesses that characterized the late nineteenth century laissez-faire era.[15] Cutthroat competition, in turn, was embedded in its own metaphorical structure, one informed by notions of evolutionary competition, or natural selection, which justified the survival of the fittest as the mechanism of progress. Competition was endemic, necessary, inevitable. The words of Tenneyson—"nature red in tooth and claw"—were frequently used to give emotional body to the savagery that pervaded a competitive society.[16] *America's Competitive Challenge* recreates this imagery in the emotional tone it invokes to justify the inevitability of our entry into the global market and the sacrifices that we must make to succeed.

The report is also rich in images that convey our fate if we are not able to compete. At the end of one section, we are "lagging," "slow," "continuing [our] decline," facing "falling competitiveness." In another, the state of our industrial infrastructure, private and public, "does not bode well" for the future; it is "worn out, obsolete, operating at full capacity." In yet another, our workforce is "poorly prepared," "ill-trained," "problem drinkers," with untold numbers suffering from "drug related problems." The implication is that if we are not able to compete in the international arena, we will slip into the abyss, witness the decay of our material infrastructure, and watch our workforce disintegrate. In short, our standard of living will fall far below that to which we are accustomed. Fear, then, is used to move us to take up the competitive challenge.

To engage in fierce competition in the global arena, there must be domestic consensus, contained in a framework that recognizes that gaining ground in international markets is paramount. To this end, we are asked to establish "the same national consensus that allowed the United States to land men on the moon, and . . . the same national effort that enabled the U.S. to establish world leadership in such industries as computers and information processing, telecommunications, biotechnology and commercial aircraft." Consensus around business aims is made credible by the promise inherent in technology, which still allows the possibility of endless prosperity without asking any group to make permanent sacrifices. Consensus revolves around "exploit[ing] the enormous pool of unused new technologies" which are "powerful tools" able to "inject new life into its [U.S.] industrial structure."

To reap the benefits of consensus at home and competition abroad, difficulties in formulating public policy have to be overcome. At present, we are burdened with "disordered public policy," characterized by "ad hocism," "administrative gridlock," and "inflexible institutions." These problems have resulted in a "failure to harness information and knowledge for productive action." If the nation's CEOs give their attention to the problem and exercise leadership, we can, through acts of individual and national will, as well as a hierarchical ordering of institutions, overcome these obstacles. In other words, what is needed is a president committed to business success who can manipulate the democratic process to create a centralized institutional structure that serves international competition.

The ideology advanced in *America's Competitive Challenge* emphasizes the primacy of the private sector and the importance of economic competition. These emphases are fostered through deregulation, redistribution, and privatization, all of which increase the private sector's power and command of resources. The success of the private sector depends on popular recognition of its centrality and reordering of the state so that global profit making is organized at the behest of American-based multinational corporations. It is a profoundly conservatively ideology, in which the power of the state and the corporate community are legitimately brought together, leaving little room for alternatives.

The disposition of societal resources implied by the tasks assigned to the several social sectors follows the ideology of the report. Each sector has its specific responsibilities. Business must learn to evaluate profits in a way that looks farther than quarterly statements. Manufacturing for export must be emphasized, together with career-long professional development for industrial, scientific, and engineering personnel. Business has a responsibility to support mathematics and science education, and should tighten its bonds with the university in areas of mutual interest—equipment, consultants, and research. Business should support university research, especially in neglected nonproprietary areas.

Universities have to develop a range of university-industry partnerships, train new professionals across a variety of fields to work for corporations in global factories and markets, and develop curricula that meet the needs of multinationals. Industry-university partnership should encourage problem-oriented research, with a special focus on how to accelerate commercialization. They should also work with government to develop data on technological trends that can be used by business and for policy making.

The workforce must be ready to retrain in high-technology areas.[17] For this to be accomplished, programs must be developed to reskill the unemployed, assist displaced workers to find jobs, and help the functionally illiterate to acquire skills for entry-level training. To facilitate the retraining of the American workforce, all students should be given a basic foundation in mathematics, science, and technology.

The disposition of societal resources necessary to accomplish these tasks is sometimes clearly specified and at other times can be readily deduced. In general, business supports those educational programs which meet its personnel training and research needs. But business will not pay the major share of these costs. At present, business pays for about 9 percent of of basic research at universities, and does not intend to pay more.[18] Instead, government is expected to provide 10 percent on top of the 67 percent it already supplies, and to further finance research through a series of tax deductions.

Universities are to redirect their enterprise—research, teaching, and service—to help corporations compete in global markets. In effect, every program becomes an M.B.A., training a new breed of multinational managers knowledgeable about high technology and its manufacture, multilingual, versed in the cultural, legal, and trade practices of the several countries in which they might work, and able to move easily between private and public sectors. The research at the heart of these programs is profit oriented and geared toward rapid commercialization. Workers and government are asked to finance the largest share for remedial education, reskilling, and training. Basic skills and remedial programs seem to be the concern of government, and workers are asked to participate in an Individual Training Account, parallel to Individual Retirement Accounts, through which they would self-finance their own reeducation when faced with the periodic dislocation likely to accompany a high technology economy.

Overall, the taxpayers make the greatest contribution to Forum initiatives, with a greater share of their monies going to government programs, largely in education. Existing institutions are asked to pick up a big share of the costs, redirecting their resources to achieve Forum goals. Since many of the initiatives identified by the Forum involve educational institutions, this also calls for increased government funding, whether through direct contributions to programs or through a variety of indirect tax subsidies. The shift in resources is justified by the assertion that in the long run all sectors of society depend on a strong economy.

Although the report speaks of a new industrial revolution that will change a traditional industrial economy to a technology-intensive economy, a large part of that transformation has already occurred. The key to understanding the revolutionary character of this report is comprehending that the companies able to compete successfully in the global market are multinational corporations. In effect, the report is asking for government subsidy as well as public and academic support for integrating the U.S. economy into an international economy dominated not by nations but by multinationals. It is not clear how traditional national interests will fare in a new international economy, where it is capital, rather than labor, that has no country.

Corporate and Campus Cooperation

In the second report put out by the Business-Higher Education Forum, virtually every text inch of its 23 pages speaks to problems faced by the university or to ways in which business and higher education are able to cooperate. The concerns of the higher learning are at the forefront. However, *Corporate and Campus Cooperation* is careful to place the partnership between business and higher education within the framework of *America's Competitive Challenge*, beginning with a reaffirmation and summary of the initial report's emphasis on the primacy of the private sector and the centrality of competition.

The structure of the argument in *Corporate and Campus Cooperation* centers formally on the question of how to maintain technological leadership in an era of competition. Yet, after this question is posed, little attention is paid to it. Instead, the report deals with why business and higher education should cooperate with each other, and what each gains from such collaboration. The report seems to have as its primary purpose the establishment of a basis for industry-university cooperation, perhaps in an effort to compensate for the lack of attention to this question in *America's Competitive Challenge*.

The independence of each sector is asserted, and their separate roles described. In the interplay of corporate and campus resources and personnel, business is interested in innovation, implementation, and development. Higher education preserves knowledge from the past, creates new knowledge, gives breadth to business, and assesses values. Although each is separate, they are complementary and interdependent.

However, the equilibrium between the two sectors has been disturbed by the problems faced by the university. Higher education currently confronts several crises: first and foremost is the fiscal crisis of the state, followed closely by a demographic crisis, which is exacerbated by regional crises. These crises have created shortfalls in operating income, which are threatening the quality of research and education. Given the problems they face, universities have no recourse but the corporation.

To deal with these problems, *Corporate and Campus Cooperation*, like *America's Competitive Challenge*, emphasizes a single recommendation: "The primary recommendation of this report is that all business corporations and institutions of higher education examine and develop means to cooperate with and assist each other."[19] The remainder of the report deals at length with the multiplicity of ways in which corporations are benefited by universities and how higher education depends on the business community. Generally, corporations are portrayed as providing resources, ranging from careers to unrestricted endowments, in return for which universities train students, produce research, and redirect their enterprise to serve the corporation.

There are three major themes in *Corporate and Campus Cooperation*. First, the report offers an antirevisionist interpretation of the relations between corporations and higher education. Second, an exchange theory is elaborated to explain this interpretation. Finally, the report weaves together the interests of the two sectors, business and higher education, in a way that creates an almost seamless cloth.

The report is antirevisionist in that it presents a history of the relationship between corporation and campus that relentlessly ignores the deep-rooted and sometimes explosive tensions between the two sectors. Indeed, the major portion of the report is titled *"Traditional Business–Higher Education Relationships"* (italics mine). Certainly the exchanges celebrated in this section occurred and continue, but no mention is made of the struggles to distance the corporation from the campus which were a part of the Progressive Era, the 1930s, and the 1960s. Nor is attention paid to the deep-rooted tradition within the academy devoted to criticizing these exchanges, which runs from Thorstein Veblen's *The Higher Learning in America* to Barbara Ann Scott's *Crises Management in American Higher Education*, with a fair amount in between.[20] Instead, very close relationships between business and higher education are presented as virtually problem free.[21]

These relationships are seen as resting on a series of mutually beneficial exchanges. The gamut of exchanges is presented in excruciating detail: corporate financial support of colleges and universities; corporate financial support of students; cooperative education; corporate associate and affiliate programs; research agreements; training programs; personnel exchanges; conferences, colloquia, and symposia; consultancies,lectureships, and faculty loans; trusteeships, directorships, and advisory services; corporate access to university resources; corporate recruiting of students; and joint projects to address national problems, such as the Business–Higher Education Forum itself. For each of these exchanges, the mutual benefits are carefully spelled out. For example, corporate financial support of students "is particularly important to higher education in an era of declining federal support." Giving money to universities benefits corporations because it "helps to enlarge the pool of college graduates that is so important as a source of new employees, it contributes importantly to good employee relations and it reinforces the role of the corporation as concerned citizen."

The interweaving of corporation and campus interests at every point occurs throughout the report. Most emphasized are training and personnel exchanges. Corporate support for students who are potential employees is stressed often and directly: for example, "The importance of the recruiting relationship is a prime motivation for corporate support of higher education." Personnel exchanges, ranging from corporate and university officials sitting on each other's boards to programs for the exchange of scientists, also receive a great deal of attention. The "cross-fertilization" provided by these interchanges is lauded, as is the window some exchanges open on university research.

However, relatively little space is given to a consideration of research activity between corporations and universities. There is no extended discussion of joint ventures or the cooperative performance of research by industry and university. In the rather brief discussion of research, contract research, in which the university does research for corporations for a fee, is identified as most important, both because it provides "high-quality research at relatively low cost" and because it is useful "in bridging the gap between 'town' and 'gown,'" increasing the interaction of personnel and thus a general understanding of the myriad interests shared by business and higher education. The university is apparently not seen as the crucial sector for basic research essential to high technology breakthroughs.

The themes stress a traditional rendition of the relation between business and higher education. Corporations provide resources, the most important of which may be careers for graduates, in return for which the higher learning prepares students for the corporate world. Not explicitly acknowledged, but implicit throughout, is the notion that universities' programs have to articulate with the economic structure if they are to be viable, and must position their students for the increasing number of promising, entry-level professional jobs located in the private sector.

Three metaphors give substance to the report's rhetorical structure. These are the "crisis" in higher education, "interdependence" between sectors, and the "vital margin" provided by corporations.

As the business community faces the global peril of competition, so the university must confront the fiscal crisis of the state. The crises are linked and are described in language that indicates forces almost beyond control: "massive trade deficits"; the need for "massive investment" in higher education to make up for "years of neglect"; a "tidal wave of students" that is now receding, presumably leaving the wreckage of the university on the shores of society. As in the initial report, this one is ridden with images of degeneration: "retrenchment," "decline," "erosion." Moreover, there is little promise of relief: "prospects for the future ... are grim"; the "picture is clouded"; "the current difficulties are but a prelude to one of the most difficult periods in its [higher education's] history"; "the need for expanded corporate aid to higher education is today more critical than at any time since World War II."

As the use of "crisis" in *Corporate and Campus Cooperation* parallels the uses of "decline" in *America's Competitive Challenge*, so their images of "interdependence" are parallel. In *Corporate and Campus Cooperation,* it is necessary to create consensus at home to avoid international economic crisis, and it is critical to bring corporation and campus together to ward off domestic crises. As indicated already, a vision of the interweaving of the two sectors is projected. It is accomplished through using images such as "interdependence," "mutual welfare," "joint dependency," "reciprocity," "strengthening appropriate ties," "expanding ties in productive ways," "interplay," and "interaction" to describe relations between the two sectors.

The corporation is tentatively presented as able to change this dismal picture. Although corporate contributions to higher education operating expenses are relatively small, they "provide the vital margin between mediocrity and excellence." Corporate support is an "important ingredient," the critical edge. What it supports is "quality," "diversity," and "independence," values long venerated by university leaders, the "hallmarks" of American higher education. Corporations are what save universities from being something more than state agencies subject to the heavy hand of public bureaucracies and the unpredictable forces of popular political currents.

The "interdependence" metaphor suggests exchange theory, which implies a trade between two partners of similar stature and does not address the question of the power of the partners. The "vital margin" imagery suggests dependency theory, which says that one of the partners engaged in an exchange needs the other, that the power relations between them are unequal, and that therefore one is able to dominate.[22] Throughout the report there is the veiled suggestion that higher education needs corporate support, despite the readily acknowledged fact that corporations provide only a fraction of the overall university operating budgets. Government is presented as unable to provide adequately for higher education, and business is presented as holding the marginal resources that matter to institutions concerned with quality. Quality involves a commitment to sustain the values of the private sector, in which the academy is always located. These values are quality, diversity and independence. As the private sector is the leaven of American economic life, so independent universities are the yeast of postsecondary education. In the struggle to sustain these values, the university depends on the corporate sector. Unless these values are sustained, both corporation and campus will lose their competitive edge.

The ideology advanced emphasizes the primacy of the private sector, in which all research universities are included. There is a strong suggestion that the campus depends on the corporation to sustain private sector values such as quality, diversity, and independence. The role these private sector institutions play at the leading edge of change justifies their institutional autonomy, the decision-making prerogatives of their managers, and their right to enormous public subsidies, with little public oversight.

Paradoxically, the corporate sector's commitment to excellence does not entail a greater commitment to providing resources for higher education:

> Higher education institutions must recognize that business corporations, unlike foundations, are not created for the purpose of making contributions to education and other non-profit sectors of society. They exist primarily to produce goods and services, and in order to prosper and grow, they must earn profits for their shareholders; thus, their contributions are subject to the conditions imposed by the marketplace and the economic climate at large.

And again, at the report's conclusion: "The need for increased investment in education must be balanced against the need for increased investment in productive capacity. No significant overall rise in corporate support of higher education can be expected until there has been a material improvement in corporate earnings." When it comes to the disposition of societal resources, corporations hold out the promise of future gifts, unrestricted as well as carefully directed, after national economic recovery. In the interim, government, at both the state and federal levels, must continue to carry the burden.

Corporations, which, according to data presented in the report, give only 1.3 percent of college and university budgets, are presented as critical to maintaining excellence in higher education. Yet traditional relations—primarily training—are stressed in the report, rather than research or innovative teaching, which might possibly make plausible the claim that a small amount of seed money will make the difference between mediocrity and excellence. Perhaps even more important than excellence is legitimation of the root values shared by the corporations and higher education, values which justify their autonomy as well as their requests for increased public subsidy.

Conclusion

The initial report, *America's Competitive Challenge*, emphasizes creating a framework of ideas and institutions that stress successful competition in world markets and calls for a wide range of support—symbolic, political, material, fiscal—to multinational corporations. In effect, it asks for government subsidy as well as public and academic support for integrating the U.S. economy into an international economy dominated not by nations but by multinationals. The second report, *Corporate and Campus Cooperation*, specifies the relationship between corporations and universities in this enterprise. Although the relationship is presented as reciprocal, corporations actually dominate, since the university is asked to direct its energy to corporate ends. Universities are expected to create human capital able to increase accumulation in the corporate sector, to engage in research that meets entrepreneurial needs, and to share faculty expertise with multinationals. In return, universities can expect careers for their graduates, a stake in the hoped-for prosperity that business-higher education partnerships will create, and honorary membership in the private sector, together with a share in the privileges that come through claiming values such as quality, diversity, and independence. Perhaps most importantly, the shared values of corporation and campus promise a political alliance in which the two sectors will work together to increase public subsidy to the private sector, very broadly construed.

CHAPTER 7

The Business–Higher Education Forum Reports: New Issues and Traditional Alliances

In this chapter, I look at the remaining four Business–Higher Education Forum reports. These are *The New Manufacturing: America's Race to Automate*; *Space: America's New Competitive Frontier*; *Export Controls: The Need to Balance National Objectives*; and *An Action Agenda for American Competitiveness*.[1] I use the same methodology as that described in the preceding chapter.

Overall, I argue that a content analysis of these reports reveals that business concerns heavily dominate the Forum's agenda, that corporate prosperity is routinely put at the top of policy agendas, and that the several reports speak not to research and production partnerships between business and higher education but to a global corporate development strategy centering on multinational market control. Ideologically, the reports call for a conservative consensus that will enable American-based multinationals to continue dominating international markets in the same way they did when American power was at its zenith, in the years immediately following World War II. To achieve this end, science is asked to focus on problems central to economic development, universities are asked to transform their graduate schools into broad-gauged M.B.A. programs, and ordinary Americans are asked to make sacrifices in terms of wages, jobs, and economic security.

In terms of maintaining the relationship between corporation and campus, the corporate community becomes the universities' political ally, pressuring the state for increased monies for research universities and a regulatory climate that recognizes their distinct organizational purpose and mission. In return, universities are expected to continue their time-honored functions of training personnel for science-based industry and legitimating corporate policy—in this case, a global economic development strategy. Although there is some concern with developing high technology research and production partnerships, it is quite limited.

The New Manufacturing: America's Race To Automate

In *The New Manufacturing*, the Business-Higher Education Forum's third report, the primary concern is once again industry. Of 38 pages, approximately 323 text inches, 14.75 inches were devoted to higher education. Of those 14.75 inches, 13 have appeared before, because *The New Manufacturing* offers essentially the same recommendations as *America's Competitive Challenge*.

As with *Corporate and Campus Cooperation*, *The New Manufacturing* reaffirms the basic positions articulated in *America's Competitive Challenge*. *The New Manufacturing* takes up where *America's Competitive Challenge* leaves off, setting a specific goal which, if reached, will make us competitive in global markets. That goal is the accelerated development of flexible manufacturing, or computer-integrated manufacturing (CIM) systems. The report is structured around making us see this as the logical next step in meeting the competitive challenge.

The report offers a number of premises which lead to the conclusion that CIM is our only recourse. The report argues that continued domestic manufacturing is essential to our economic well-being. It makes the case that in the global market there are three essential requirements for successful manufacturing—quality, cost-effectiveness, and the flexibility to respond rapidly to changes in market demand. Traditional modes of manufacturing—mass, batch, and custom—cannot meet these criterion. Therefore, we have to accelerate our development of the factory of the future. The new manufacturing will use advanced forms of flexible automation, integrating all functions—design, engineering, materials handling, fabrication, assembly, inspection and testing, and sales and distribution—through computer and communications technology.

The major themes in the report are the centrality of manufacturing; the possibilities of the factory of the future; and the changing structure of the workforce.

The centrality of manufacturing is treated in terms of competition in the global market. As traditional industries mature, the United States loses technological advantage because standardized manufacturing processes can easily be duplicated by other nations. If current trends continue, standardized commodities will not be produced in the United States. Yet manufacturing is critical to a vital economy. The value added by productivity gains in the new manufacturing is what will generate substitute service-sector jobs.

In stressing the centrality of manufacturing, the report reflects the composition of the majority of the corporate members of the Business–Higher Education Forum, most of whom head manufacturing firms.[2] Although the report takes the socially responsible position that it is committed to domestic manufacturing, in point of fact most of the corporations that are signatories to the report have a strong multinational presence, which involves offshore manufacturing. It is not clear that they are prepared to restructure their current operations, although there is an implication that, given the government subsidies proposed in the recommendations, they would build factories of the future at home. If the worker-free CIM factory were realized, there would be no cost benefits from lower wages offshore, but neither would there be a significant increase in domestic jobs.

The factory of the future is dwelt upon in detail. At its heart is the computer, not the robot. The aim is integrated computer-controlled production, with little or no human intervention. Reprogrammable CIM will continuously produce specialized commodities on demand. Its applications will eventually extend beyond the factory walls to sales, distribution, and delivery. The major cost and management gains achieved by CIM are lower labor costs, elimination of the unpredictable human factor, and flexible machines able to be retooled by computer for new tasks.

Accelerated use of the new manufacturing will cause changes in the structure of the workforce. There will be short-term transitional problems, where displaced workers will have to be retrained, but in the long run increased productivity will generate more jobs. Little hope is held out for any increase in production-line manufacturing jobs; instead, the report emphasizes the creation of service-sector jobs, or service jobs within manufacturing. The types of service jobs that will emerge, as well as their pay and benefit scales, are not directly addressed. Although not stressed, the possibility of a two-tier job structure, with an expanded executive class and a mass of service workers, seems likely, given the labor market forecasts used and the organizational changes projected within firms.

The rhetorical structure of the report clarifies its thematic content, fleshing out the bare bones of economic change with metaphor, image, and analogy. The metaphor of competition continues, captured by the "race to automate" in the report's title, but it is not central.[3] The dominant metaphors are the factory of the future, the displacement of workers, machines as humans, humans as problem solvers, and management coping with new organizational structure.[4]

The "factory of the future" is aptly described with the language of systems management, the behavioralism of business schools, a discourse which objectifies and simplifies human interaction as well as individuals' grasp of complex processes. The images of the production process emphasize its lack of problems, human or otherwise. The report sees the factory as a site characterized by "ease of manufacture"; "simplifications in design"; computers that "plan," "design," "integrate," and "make sets of instructions" for robots; that work during "unmanned ... night shifts."[5] The problems of automation are treated in three short paragraphs that dwell on human limitations rather than those of machines.

Although the new manufacturing may bring great gains in productivity, enabling the United States to compete in the global marketplace, the factory of the future has been announced as immediately impending at regular intervals since the 1940s. Moreover, the history of automation is rife with problems, ranging from defective equipment and underutilized machinery to labor struggles that move up the occupational hierarchy as low-level jobs are closed out.[6] It seems unlikely that operationalizing the factory of the future will be as problem free as the imagery suggests.

In an analogy used several times, the hope is raised that the new manufacturing will follow the path blazed by agriculture. An economist is quoted as saying, "If American manufacturing is going to continue as a strong contributor to the economy, it must go the route of American agriculture: that is, fewer workers, greater productivity, higher volumes of production, and improved quality." Nebraska is the agricultural state we are asked to emulate. "Fewer than four percent of workers ... are directly engaged in the basic business of farming, but more than one-third of ... total workers are employed in supporting businesses." Neither the out-migration from the state nor the average service-sector wages are discussed.

If American agriculture points to the promise of the new manufacturing, it also raises a number of questions not discussed in the report, most of which turn on who benefits. American agriculture is marked by increased concentration of ownership, production that emphasizes costly commodities high in the food chain, and reliance on petrochemical fertilizers and pesticides that frequently create environmental and nutritional problems.[7] Agriculture is currently at the forefront of technological innovation, drawing on biotechnology, particularly in seed development. Evidence thus far indicates that the use of biotechnology will follow the pattern established by corn hybridization, where seed is exploited as an annual cash commodity tightly controlled by corporations rather than as a renewable resource handled on the family farm.[8] And in view of present worldwide grain gluts, the agriculture analogy raises the possibility of advanced techniques resulting in overproduction rather than increased profits.

The "displacement" of workers is a euphemism for unemployment, a word not used anywhere in the report. "Displacement" is discussed in terms of the entire U.S. manufacturing system, so the "sacrifice" and suffering of individuals and regions are not dwelt upon. In this parlance, "tens of thousands" of workers losing their jobs is a "relatively small number." Predictions as to how many jobs will be lost if the move to the manufacturing is "accelerated" are not clear. The report is clear that the new jobs available for displaced workers are likely to be in the service sector. However, the type of job, the specific skills needed, and the level of remuneration are not explored. Instead, the retaining of displaced workers is treated in the context of how humans will fit into the automated factory. Machines are treated as persons: they have "intelligence," "sensory information ... systems" "vision," and can "read." Humans will have to compete with machines for jobs, and will do best if they concentrate on skills that machines do not have particularly "their innate human capabilities to deal with a wide spectrum of problem solving situations." No doubt displaced workers can draw on these capabilities when, during their labors at the frontier of the service sector, the fast food franchise, they are confronted with disgruntled customers eating high technology hamburgers, harvested at rapid rate through calf embryo transplants, and processed in automated animal disassembly plants.

The central task of management will be "to rethink the fundamental roles of, and relationships between, humans and machines in the factories of the future." Managers will have to change from organizing and supervising "large groups of people in a single plant" to supervising "plants that might not be manned [sic] at all." That this "reality argues for a new view of labor relations," seems something of an understatement. Organizational change is expected to go farther than the factory floor, with "intelligent" machines reducing middle management and "flattening" hierarchical structures. "Executives [will] become more intimately involved in the production process itself," and when they do deal with colleagues and workers they will be increasingly likely to do so via computer terminals "rather than face to face." Overall, the images are of managers finally free of labor problems, from production workers to middle management, distanced from routine wear and tear of human contact by their sophisticated machinery, yet in even greater control.

The ideology of the report emphasizes the benefits of a production process that deals with labor costs and problems by eliminating workers. The majority of the population is consigned to the service sector, which is peripheral to manufacturing and not fully discussed. The values stressed are profit and control.

The report concludes by arguing that the factory of the future will not necessarily increase oligopoly, since advanced manufacturing systems should theoretically allow small firms to be more competitive, provided they can manage the entry costs. However, given the data presented earlier in the report on the enormity of start-up costs, it seems unlikely that small firms will be able to enter. The section treating market structure appears abruptly at the end of the report and may be tactically necessary, but it is not particularly convincing.

The deposition of societal resources follows the trend set in *America's Competitive Challenge*. Indeed, the recommendations made by *The New Manufacturing* are recycled, in barely altered form, from *America's Competitive Challenge*. The primary objective is adjusted to fit the report's topic: "The achievement of the primary objective—improving American manufacturers' ability to compete through the accelerated use of advanced flexible automation systems—will require concerted efforts by all sectors of society: government, business, educational institutions and labor." As in *America's Competitive Challenge*, the taxpayer pays the major share and government is in charge of the major initiatives, but in *The New Manufacturing* labor more clearly bears the burden.

The *New Manufacturing* raises as many policy problems as it proposes solutions. Foremost among them are the national responsibilities of multinational corporations: how do corporations engaged in international manufacturing participate without prejudice in the debate about their domestic role? Another problem is the priority given to automation. Finally, there is the question of how mass education in science and technology, repeatedly stressed, fits with what seem to be proposals for two-tier job structure in which there is actually a very small demand for highly educated people.

Space: America's New Competitive Frontier

The fourth Forum report, *Space*, was approved by the Forum for publication four days before the Challenger space shuttle blew up. The Forum decided to release the report despite the tragedy. The Executive Committee took the following position: "Consistent with President Reagan's request that the nation continue its pursuit of space initiatives, we have decided to publish our report as scheduled."[9]

The report is doubled columned, with approximately 50 pages of nonrepetitive content, or roughly 800 column text inches. It is the longest of the several reports. Of this material, approximately 94.5 column inches of text are devoted to the higher learning. The ratio of educational to other concerns, then, is 1:8.5.

The central question with which this report deals is how to create the conditions for commercial development of space. It assumes that the competitive challenge outlined in earlier reports extends to space. It further assumes that government, business, and higher education will work in partnership to further conditions for commercial exploitation.

To advance the process of commercial exploitation, the report recommends that government, business, and higher education work to bring together a constituency that will support space initiatives, create a policy formation network that will foster space development, and build public infrastructure for extraterrestrial private development. Critical to these several endeavors are a continued flow of government resources, clear government policies that regulate space in ways beneficial to commercial growth, and concrete incentives for business. The incentives usually take the form of subsidies from which both business and higher education benefit—expanded research funds, increased automation efforts, and the rapid development of a permanent space station.

Although the structure of the argument is relatively straightforward, the themes reveal complexity and ambivalence. The major themes are three: the promise of commercial development of space, the possibilities that space presents for international relations, and the potential of research. Each of the major themes expresses a positive thrust but also touches on and sometimes explores the problematic aspects of development.

The rapid commercial development of space is presented as critical to the future health of the U.S. economy, as is the expansion of private investment and involvement. Economic innovations that have already spun off from the NASA program are elaborately detailed as evidence of the symbiotic relationship between space and economic development. Future benefits from the space industry are spelled out: new jobs, lower product costs, and improved balance of trade. However, even as the gains of private enterprise are extolled, the report is uncompromising in insisting that the federal government will have to continue to bear the costs of development for the foreseeable future. The risks are so great, the likelihood of a near return on investment so small, and the overall costs so enormous that continued government involvement is imperative. Although privatization of government space infrastructure is advocated as essential to commercial development, the public will have to subsidize these efforts for the foreseeable future.

The special problems space poses for international relations are addressed. Space is seen as an arena in which to promote international peace and cooperation, but also vividly presents possibilities for competition. These competitive threats are both commercial and military. On the one hand, there are fears that cooperation will result in other countries copying our space technology and being able to market it more successfully because of lower labor costs. On the other hand, there is a strong sense that progress in space has been made through enduring competition with the Soviet Union for strategic and diplomatic ends. Yet even as threats are articulated, the need for cooperation is stressed: "As a world leader, the United States should accelerate its efforts to explore mechanisms for international collaboration on the peaceful uses of space." The Forum understands these positions as contradictory and asks for a resolution that will enable the United States "to cooperate with other space nations without endangering any competitive advantage."

Realizing the potential of research is seen as essential to the commercial development of space. But maintaining an appropriate balance between applied and basic research is problematic. The present development stage of the space program calls for a great deal of applied work, yet basic research, usually defined as separate from development, has traditionally been viewed as essential to the major scientific discoveries on which long-term technological progress turns. The dilemma is resolved by calling on science to engage in basic research that has immediate potential for commercial application.

The themes in *Space* recognize the problems as well as the possibilities of space commercialization. The central problem is cost, on which depends not only commercialization, but also space development for strategic and diplomatic use as well as for research. The immense costs of continued government involvement create ideological as well as economic costs. Rather than presenting the private economy as separate and better than the public, the report acknowledges the interdependence of the two, not for a temporary "bail out" or for special "targets," but for the foreseeable future.

The rhetorical structure is rich in metaphors of competition, echoing earlier reports, but the central metaphor is that of "partnership." The partners in the commercialization of space are government, business, and higher education. These institutions are in a "partnership for progress" in which there is an "evolution of leadership roles." The language used suggests that their growth is part of a social process of steady improvement and natural maturation, not the product of political choices made by an engaged citizenry. Business and higher education respond to these evolutionary social forces by "accepting an increasing share of the responsibility." Although the partnership is a natural development, industry and education are the active agents. Together, they work at "leveraging accomplishments into benefits for the American people."

Although partnership means "joint interest," very different imagery is used to describe the parts played by the several partners. The rhetorical structure used to describe the role of the state in the commercialization of space is very uneven, evidencing both dependence and repugnance. Although government will continue to be the "dominant force," the political process is described as unpredictable and chaotic, something to be controlled and contained. For example, "the election cycle and the annual budget battle foment constant struggle and make continuous, rational planning virtually impossible." Similarly, the cost of federal programs, although understood as essential to the exploitation of space, are seen as unmanageable. "Each new long-life system has its own operations and maintenance needs stacked on top of those earlier systems, with little relief from customer revenues." The "privatization" of state resources is advocated at the same time that the distant day when "the private sector ... assumes major funding" is longed for. The inconsistent imagery probably reflects the difficulty representatives of the private sector have in thinking through policy in an age when the state is ever more essential to capital formation.

The part played by industry in the partnership is described with language that makes business the prime mover. The image of "leveraging," with its connotation of making manifold gains through the knowledgeable manipulation of ordinary elements, appears frequently in the report and emphasizes the transformative powers attributed to the private sector. Companies are seen as bringing to the partnership an "aggressive but rational approach." They "maximize benefits" and "identify real market needs." They take risks, "staking their future on extraterrestrial ventures." Business leaders are implored to use their "vision" to transform space into successful commercial enterprises. As the private sector provided the "vital margin" in earlier reports, here too it is the active agent.

The metaphor for higher education's role in the partnership is that of a private sector institution that should "promote economic growth." Higher education is presented "as a natural partner to business... [with] a high stake in the future use of space for commercial purposes." I take this role to be metaphorical—a figure of speech in which one thing is spoken of as if it were another—in that higher education is not private in the same sense in which a profit-maximizing corporation is and in that the promotion of economic growth is not an educational end about which there is agreement either inside or outside the academy. Metaphors such as this obscure the taxpayer's extensive subsidy of public and nominally private institutions and ignore criticism that speaks against using higher education to further private economic goals.

The model invoked to suggest higher education's participation in space development is the Morrill Act. As the Morrill Act was used to push back the frontier and create the infrastructure for industrial agriculture, so a "Space Grant University" is pictured as providing the "intellectual foundations... to build a dynamic new industry and fulfill a national commitment." The Morrill Act conjures up images of pragmatism, progressivism, and research that serves the public good, often through agricultural development. What is not examined is how such legislation was used to manipulate the political process and to whom social and economic benefits flowed. According to some students of the land grant institutions, the federal bureaucracy, together with the extension service and universities, initially used their organizations to engage politically conservative farmers on middle-sized farms, cutting out small farmers, tenant farmers, and sharecroppers,. Eventually, these same forms of organization came to serve large corporate farmers rather than middle-sized farmers.[10]

The modern models of "public-private collaboration on the state level" presented in the report suggest that questions about who benefits are still salient. The Indiana model, held up for emulation, consists of a state Corporation for Science and Technology, which is composed of presidents of Indiana's research universities, the CEOs of leading state corporations, the state superintendent of schools, and one representative from organized labor. "With funding provided by the state, the corporation considers research proposals, which all must explain how the basic research can lead to a commercial product or process." In this model, university administrators and presidents of private corporations decide how public money should be spent. The criteria for making decisions is the potential basic research has for commercial development. Public participation is minimal, and faculty do not participate at all. Traditionally, the coupling of basic research and commercial development would have been read as a contradiction in terms.

As was the case with previous reports, the ideology advanced stresses a public-private partnership in which the private sector uses the state to administer and subsidize its agenda. The differences between the state and the private sector are minimized by calls to collective actions to achieve urgent national goals, in this case the commercialization of space. Although universities are included in the private sector, they, like the state, are expected to reorient their goals to meet the needs of corporations engaged in the development of the space station.

The disposition of societal resources follows the ideology being developed. Essentially, government is to pay for the research and development necessary to commercialize space, and to work to broaden public support for the space program. Colleges and universities are to work on federally funded space programs; participate in joint ventures between government, business, and universities; broaden core curricula to incorporate space age concerns; and develop training programs in areas ranging from telecommunications technologies at community colleges to space law at universities. Business is asked to "take advantage," to "encourage and support," to "investigate," and to devote "more emphasis" to space technologies, but is not asked to commit significant amounts of capital.

This report moves the Forum beyond the transformation of a national economy to a global one, bringing us to the galactic. For a venture of this magnitude to get off the ground, the state has to be involved on a number of levels: it must create a political and bureaucratic constituency for extraterrestrial development, contribute heavily to capital formation, and make the expected contributions to the intellectual and material infrastructure. Industry and academe make the development decisions, while the state organizes their efforts and supplies resources.

Export Controls: The Need to Balance National Objectives

The fifth Forum report, *Export Controls*, deals with administrative controls imposed on the export of military, commercial, and scientific products and processes as well as of data. The report is approximately 296 text inches, of which 27 are devoted to higher education, if the topic of scientific communications is understood as subsumed by higher education. The ratio of higher-education to business concerns in the report, then, is 1:11.

The policy problem posed by the report is how to establish the proper balance between increasing export controls in the name of national security, on the one hand, and easing export controls so as to expand our share of world trade, on the other.

The structure of the argument assumes a permanent state of Cold War. This presupposition means that some form of export control which prevents the enhancement of the Soviet military capability is necessary. These assumptions frame the policy problem and shape the answers given. The ability to deal creatively with these problems is further limited by the recognition that European countries do not express the same need to erect barriers to the free flow of global trade as does the United States.

To preserve these paradoxes, the development of a narrow set of short-term militarily strategic goods and data for control is advocated. These products and processes should be accompanied by a "decontrol" schedule. With the exception of this limited list, controls should no longer be unilaterally imposed by the United States but should be multilaterally enacted by CoCom [Coordinating Committee for Multilateral Export Controls], "a multilateral control group made up of NATO countries (except Iceland and Spain) and Japan."[11]

In keeping with this narrow definition of export controls, all nonclassified scientific research should be free from government restraints of any kind. Controls should not be used to achieve foreign policy objectives, for example, attempting to compel the Soviet Union to abandon Afghanistan. Such efforts are by and large ineffective and are not supported by our CoCom allies.

To realize these positions, a more efficient and accountable administrative system of controls is needed, with sufficient resources to ensure that the system works. The power of the executive branch, particularly the President, to impose export controls, especially with regard to obtaining foreign policy objectives, should be curtailed. Instead, that power should be closely overseen by Congress.

All in all, the report is aimed at curbing the power of the executive, especially with regard to the growth of a national security state. However, in assuming a permanent state of cold war, the Forum concedes the necessity for the state apparatus it seeks to restrain. At best, a precarious balance is struck, where commercial global imperatives are weighed against military. At worst, the ideology and machinery for a national security state remain intact.

There are at least three major themes. They deal with the emerging independence of non-Soviet bloc nations, ways of figuring the calculus of costs and benefits in an international arena, and the nature of science and technology in a world where information is increasingly commodified.

The Europeans, Japan, and an increasing number of third world nations are seen as becoming influential in global markets and politics. Other countries are recognized as having the same scientific and technological capability as we do, sometimes even a superior capacity. These countries do not necessarily view the Soviet Union as an ineluctable adversary and instead seek expanded detente and trade with Russia. As a result, they are very often willing to sell the USSR state-of-the-art technology. Although we can exercise some influence over our allies, perhaps checking the flow of near-term "militarily significant" technology, we have lost the economic leverage we exercised over them in the era that followed World War II.

The report assesses these changes in terms of costs and benefits. The high costs of export controls are weighed against the likely benefits, given our inability to control our allies, and are found wanting. Even when applied unilaterally only to near-term militarily significant technology, they heavily penalize American corporations in terms of their position in world markets. When it comes to controls applied unilaterally to obtain foreign policy objectives, the costs are excessive and few gains are realized because all that can be achieved are symbolic victories, given the dynamics of world trade. Moreover, the controls themselves extract a high cost. The bureaucratic machinery is slow, ponderous, and inefficient, adding unjustifiable expenses to the cost of doing business in international markets.

Technology and science are seen as being very difficult to control strictly. Technology cannot be tightly held: its export means that eventually the secrets it contains can be grasped by scientists and engineers in other nations. Science is excepted from control on somewhat different grounds. Its excellence depends on an open system, characterized by the free exchange of information and consistent evaluation by peers in an international community.

The themes embody the slowly changing and contradictory attitudes of the Untied States to a shifting balance of world power. Because of the commercial and technological success of other nations, the United States is at best *primus inter pares*. On the one hand, the report recognizes that the United States can no longer dictate diplomatic or commercial positions to its allies; on the other, it still expects consensus on the part of its allies. Commitment to the Cold War, with its intense focus on East-West conflict, seems to make it more difficult to understand the limits of power in a new world order.

The rhetorical structure closely follows the structure of the argument and thematic content. East-West relations are described in muted Cold War language, which underscores the weaknesses of the Soviet economy and the advanced state of its military. The Soviet Union is "technologically backward," not because of a "deficiency in [the] scientific base," which is quite strong, but in terms of "developing, disseminating and applying technology and new products." In other words, the USSR is a threat militarily but not economically, a position that covertly celebrates the strength of capitalism and justifies continued hostilities between the two countries no matter what political changes occur in the Soviet Union.

West-West relations are described using language that recognizes a divergence between U.S. and CoCom policy at one level, but at another assumes a shared belief system that will reconcile differences."There ... are strong sentiments in Europe in favor of *detente*," "a lack of enthusiasm" for broad export controls, and a tendency "to resist U.S. urging to expand" controlled items. Although these differences are catalogued, the possibility that they might finally prevent cooperation is not recognized. Instead, there is an assertion of a "fundamental consensus" between the United States and CoCom centered on the shared belief that items of military importance should not be exported to the Soviets.

The control system itself is described in almost Kafkaesque terms. The bureaucracy has an "expansive reach," a "broad sweep," the "scope of controls is excessive," and the "complex matrix ... formally encompasses all the means" of export. The Militarily Critical Technologies List is the "size of the Manhattan phone directory and is so comprehensive that it has been dubbed by some 'The Modern Technologies List.'"

Most technical knowledge is treated as a commodity, either "classified," appropriated by the military, or "proprietary," owned by private corporations. Under the Reagan administration, the amount of information classified increased drastically. Both the Department of Defense and the Department of Commerce are involved in the new classification schemes. Corporate capital too is making an effort to control knowledge. On the one had, it is trying to develop and enforce international property laws. On the other, it is struggling with the state, specifically the executive branch, and particularly the Department of Defense, to allow for the free flow of state-of-the-art technology in trade.[12]

In the increasing commodification of knowledge, science is the exception. Scientific advance is depicted as depending on "the free exchange of information," the unimpeded "give-and-take" which plays an essential part in "sparking creativity and in encouraging the leaps of interaction by which major advances are often achieved." In the end, "the national security of the United States is best served by a policy of protecting the nation's scientific and technological preeminence through maintaining openness, rather than by imposing controls."

The rhetoric surrounding science is not new. It seems to be drawn from positions advanced by AAU presidents, which in turn echo the position advanced by the academic community since World War II.[13] As Howard W. Johnson, president of MIT, put it during testimony before Congress in the early 1970s, "I am sure you are aware that in areas where we do not have the classification of research we lead the world, such as computers, and in other areas where we do not have classification of research, like nuclear reactors, we do not necessarily lead the world."[14]

University presidents, past and present, seek to keep the state from encroaching on science. However, their rhetoric does not seem adequate to the task, especially when compared with the clear proposals dealing with export controls. In contrast to the detailed recommendations for reforming the mechanisms of export control, little attention is paid to working through how the free exchange of scientific knowledge will be protected once it assumes concrete shape—in journal articles, grant proposals, technical reports, and conference speeches. Although business leaders are willing to support university presidents' effort to curtail growing state classification, *Expert Controls* focuses more on facilitating the free flow of trade than on maintaining the openness of the scientific system.

The recommendations developed in the report are quite specific. The goals of the export control system should be redefined and narrowed and its bureaucracy redesigned. The economics of reform are not excessive. However, there are costs in terms of control, and these are born by those sectors of the government whose power is to some extent curtailed: the President and the DOD.

In a way, this report deals with conflict and compromise among elites. Despite tensions between a corporate community seeking increased commodification of knowledge and a scientific community trying to enhance the free flow of knowledge, corporations and campus communities have aligned themselves against the executive on the issue of export controls. Although they are in conflict with the executive, they accept the same premises: the Cold War, the superiority of free markets, the solidarity of the allies, and the right of America to global hegemony in economic and military matters. Thus, they are willing to leave the national security state intact, ensuring that issues of control of science and technology will surface over and over again.

An Action Agenda For American Competitiveness

The sixth and final report considered is *An Action Agenda for American Competitiveness*. This report is a departure from past practices for the Business-Higher Education Forum. Instead of limiting participation to industry and university presidents, it included members of Congress as well as representatives of federal and state agencies. The new participants account for roughly a quarter of the group who framed this report. The several groups tried to articulate their consensus on policy areas on which they were in agreement—research and development policies, human resource policies, and trade policies—and made rather specific recommendations, which presumably the politicians will bring before their legislative bodies.

The inclusion of new participants apparently called for a different look for the report. It is written on slick paper, with light colors, unlike the somber, heavy documents of previous years. There is less print and more pictures. Altogether, there are 196 column inches of text, of which 37 are devoted to higher education, broadly construed. The ratio of concern for higher education to other interests is 1:5, the highest yet expressed, with the exception of *Corporate and Campus Cooperation*. There is probably less direct attention devoted to business in this report than in any other.

However, the structure of the argument is almost identical to that of the initial Forum report, *America's Competitive Challenge*. The central issue is regaining corporate competitiveness in global markets. The case is again made that the American economy is suffering from serious, deep-rooted problems that cause us to lose ground to foreign competitors. The research and development infrastructure is still being dangerously eroded. Education and retraining programs continue to reveal serious problems. Government policy-making machinery needs to be overhauled and streamlined so that policies on which there is consensus can be implemented. Short-term gains have to be sacrificed for long-term benefits. We must develop a coherent national strategy to implement our policies on competitiveness.

Although the structure of the argument of *America's Competitive Challenge* and *An Action Agenda* are alike, there are clear thematic differences. In this report, the primacy of the private sector is not as boldly asserted and political leaders as well as state bureaucrats are included as decisions makers. Deregulation, redistribution, and privatization, all of which were initial themes that enhanced corporate prerogatives, are not as strongly sounded here. Instead, shifting resources within government are emphasized, whether from defense research to economic development in terms of research and development, or from social welfare to job training in terms of human resource policies.

However, there are thematic similarities. Global competitiveness is central. Consensus among elites—corporate, university, and political—is seen as essential to reaffirm and enact policies that will enhance competitiveness. And, in the final analysis, the proper work of government is still to create a "climate for competitiveness."[15]

Although international competitiveness is presented as the primary issue, few proactive proposals are developed to promote competition. Instead, methods of containing foreign encroachment on U.S. markets and products are put forward. Rather than develop economic models designed to enhance exports, as have other countries, the United States is simply exhorted to increase its access to foreign markets.

Although there is a growing realization that some sort of national planning is necessary to address international issues, this realization is accompanied by a reluctance to develop any specific mechanisms for fear that they will constrict the power of the private sector. There is a call for "a far sighted consensus 'strategy'" in which "public and private sector leaders should seek to eliminate the contradictions and obstacles in public policy and private operating procedures." At the same time, the report warns several times against choosing "a top-down, national-planning approach."

Rather than advocating the redistribution of resources from the public sector to the private, as did *America's Competitive Challenge*, *An Action Agenda* calls for changes in funding patterns within the public sector, accompanied by mission shifts. Too much money is going to defense research, and too little to basic. However, blanket accounts for basic research are not proposed. Instead, resources should be concentrated in "gray areas" where science and technology overlap, creating the possibility for commercial development. So too, human resource policies should be altered, with funds shifted from welfare, or income maintenance, to training. These shifts seem to be changes in emphasis rather than substantitive changes in priorities. Science and education receive a larger piece of the pie, but research is supposed to expand the pie magically so that it still serves as many people, and perhaps more.

The rhetorical structure of *An Action Agenda* is similar to that of *America's Competitive Challenge*, but perhaps more dramatic. The central metaphors are of decline and decay, peril and paralysis. They describe the failure of the economy and the problems this failure creates for the class structure, the research and development infrastructure, and government.

Notwithstanding recent talk about recovery, the economy continues to have "long term, deep seated problems, just below the surface." "Declines are occurring across the board." The US has "not kept pace with the gains of other nations." In some cases "US producers have simply given up," and there has been an "exodus of US production." The images used portray U.S. manufacturers and government officials as victims of external forces, and give no indication that they are in any way responsible for overseas manufacturing, tax laws touching multinational corporations, or a foreign policy that has aided and abetted our competitors as a part of long-term Cold War diplomacy.

The decline of the economy has consequences for the U.S. class structure. "America's standard of living is declining...America's middle class is shrinking...America's underclass...is growing...the ultimate victims...[are] the American people." The parallel but unarticulated subtext is that the possibility of upward mobility, the growth of the middle class, and the containment of the largely black "underclass" depends on private sector success in international markets. Because these points are made in subtext, there is no need to detail the many problems of transforming corporate prosperity into gains for the middle and underclasses.

The situation for the research and development infrastructure, which is critical to maintaining our technological edge, mirrors that of the economy. The "R & D infrastructure... is falling apart." There are "disturbing signs." "The facts and figures documenting the deteriorated state of America's university enterprise are plentiful... and grim." The emotional refrain running just beneath the surface is the same one that has most often informed requests for greatly increased funding. The rhetorical structure creates an atmosphere of impending crisis. External threats, which heighten internal weaknesses, are emphasized, and even exaggerated. Fear and competition are used to drive federal science funding.

The central image for government is "gridlock." The nation's policy-making machinery has become paralyzed by a gridlock—"ensnared in a complex maze of competing special interests, overlapping turf, and inconsistent actions." Research and development is characterized by a "series of ad hoc and haphazard policies." There is a "maze of uncoordinated federal, state and private sector training programs." "Within the government sector itself, policies are made by multiple players at the federal, state and local levels, leading inevitably to overlap, contradictions, or both." The gridlock metaphor conveys an image of overload and the consequent breakdown of critical systems. Individuals acting within the available frameworks of choice have not enhanced their own self-interest but have caused system paralysis. "Gridlock" and the attendant images of chaos implicitly call for a reordering of the system, a reinstatement of authority.

The thematic and rhetorical structures contribute to an ideology that places a high value on continued prosperity. The insecurities of the middle class—loss of status, standard of living, and opportunity for upward mobility—are played upon to create a climate in which greater order and control come to seem the only way out of crisis. As the report's recommendations indicate, the reordering of social systems involves the reassertion of private sector authority over government.

There are four sets of recommendations—for research and development, human resources, trade policy, and integrating federal policy. All call for greater governmental involvement and greater private sector authority over government systems. In terms of research and development policy, government must assume responsibility for the maintenance of the college and university research and development structure on a predictable basis, and also bear new capital costs. In terms of fostering "the swift translation of knowledge to commercial products and processes," government should improve the collection and dissemination of science and technology information worldwide and underwrite joint technology development ventures among business, academe, and government. In each of these proposals, there is a focus on creating easy access to the executive on the part of concerned parties—appoint a national competitiveness adviser, expand the responsibilities of the national security advisor to include a review of the impact of defense policy on competitiveness, form a public-private Council on Competitiveness, and reorganize the federal administration so that mechanisms are developed to create coherent competitiveness policies across agencies. The report, then, concentrates on ways to incorporate business concerns—"competitiveness"—at the apex of bureaucracy, often through the creation of additional offices, rather than on streamlining or decentralization.

Although these recommendations echo the concerns of *America's Competitive Challenge*, there are notable differences. The initial report treated research and development as the province of industry rather than of academe, and would have made significant amounts of research money available to corporate-based, high technology science. Much of the money would have been returned to corporations through a series of tax breaks. In *An Action Agenda*, recommendations are centered more on the needs of universities than of business, and subsidiaries are more often directed to the academy than to industry.

The human resource policies proposed continue to use education as the primary mechanism for ameliorating the harsh consequences of an economy in transition from a national to an international system. Government pays for programs to reduce functional illiteracy, to improve the U.S. employment service, and to reform the unemployment insurance system, and "dislocated" workers bear the primary cost of financing their retraining programs. Colleges and universities are asked to contribute by increasing the teaching of foreign languages and cultures so that "businessmen" can operate in an international milieu. The recommendations conclude with a call for more liberally educated college graduates, which contradicts earlier appeals for a broad-based education through high school and specialization at the college level. Unlike the initial report, this one pays somewhat more attention to improving the education system at its base than at its peak. However, the taxpayer still bears the cost.

The trade policies proposed are very general, perhaps reflecting the difficulties of forging agreements between representatives of the several sectors. Access to American markets by foreign companies should depend on reciprocal access by U.S. firms to foreign markets, with special attention to the Japanese. U.S. trade law remedies should be modernized to give firms relief from illegal foreign practices. U.S. intellectual property should be vigorously protected. All of these policies depend on increased vigilance by government and an extension of commercial law enforcement to international markets. Unlike in the initial report, the trade policies do not deal with the major problems suggested by the text—assistance for corporations in high technology fields, accountability of multinational corporations, national mechanisms for shaping and implementing trade policies, and establishment of international free trade.

An Action Agenda suggests that achieving consensus constrains policy initiatives. Although the report echoes the major concerns of *America's Competitive Challenge*, the insistence on the primacy of the private sector is diminished, and the voices of other political actors are heard. The corporate sector claims less in terms of direct and indirect subsidies, and government is mobilized to deal with the problems of the transition from a national to an international economy. The university may benefit most: it is perhaps easier for politicians to subsidize competitiveness through science than to help corporations directly. However, the benefits accruing to the university are concentrated on creating knowledge useful to corporations. "A top priority must be to focus on initiatives that occur in the 'gray area' where scientists, engineers and entrepreneurs draw upon the world's scientific and technical knowledge, as well as their understanding of markets to develop new products and processes that can meet important social and economic needs." The university increases its claims on government, moving from a percentage increase in the research and development budget in *America's Competitive Challenge* to maintenance and capital costs. These gains from the government bring the university into closer cooperation with corporations, since funding is tied to the utility of knowledge for economic development.

The Reports in Retrospect

The material in the preceding chapters has emphasized the similarities between themes that appear in presidents testimony before Congress, in CEOs' remarks in the public press, and the Forum documents. Equally important are points of contest and points of difference for these enable us to see which group gives ground when establishing joint positions. The material in Chapter 4, based on the presidents' testimony in Congress, provides a basis for assessing presidents independent of the Forum; Chapter 5 gives a basis for looking at the CEOs in the same way. This chapter and the preceding chapter give us a basis for comparison through an examination of material jointly produced.

Each group has their own terrain, zones of interests which are important to them in which the other group is not interested. For example, presidents are concerned with general support for higher education and student financial aid, and CEOs are not. CEOs are very interested in federal rules, regulations and support for specific industries, and presidents are not. These are areas in which there is no overlap between the two groups, not areas in which there are points of contest or points of difference. Indeed, there were very few points of contest, points on which the two groups were at odds with each other. However, there were several points of difference, points on which one group changed their position in Forum documents. These points were points of attitudes toward business and higher research partnerships, toward the state and toward ideology.

In their Congressional testimony, university leaders frequently emphasized the possibilities for universities to engage in joint research leading to shared entrepreneurial activity with corporations, whether through patent development or technology transfer. CEOs, in their remarks in the public press, did not speak specifically to joint partnerships centered on research. Although the Forum was highly supportive of universities working closely with business, it rarely spoke of research partnerships or entrepreneurial ventures, but more often addressed traditional sharing, such as contract research and training. This may not matter to university presidents, since the legislative agenda and ideology promulgated by the Forum create conditions that allow the university to negotiate independently partnerships built on entrepreneurial science.

In their testimony before Congress, university presidents always saw themselves as external to government. The presidents rather consistently argued against federal intervention on matters ranging from classification of research to personnel. Even when counseling Congress about the presidential science advisory structure, presidents were careful to recommend an advisory body that would represent the point of view of university-based scientists rather than create an organization in which professors functioned as state agents. This externality may be an artifact of the structure of the testimony, which usually handles requests for monies, thereby creating a format that automatically makes the presidents supplicants, outsiders in the corridors of power.

Although the Forum documents are jointly created by university presidents and CEOs, the CEOs posture toward the state, as reflected in their remarks in the public press, is the view that appears in the reports. Formally, the Forum makes a very sharp distinction between civil society and the state, one that valorizes civil society, but the documents speak again and again of using the state to organize corporate competition, especially at the international level. The Forum, then, takes an instrumental view toward the state, seeking to incorporate into civil society those aspects of the state that would serve the private sector.

The presidents do not see higher education as having an ideological function. In their testimony before Congress, they indicate over and over that they believe advanced education creates science, not ideology. They do not see science as having an ideological component that might generate broad student or public support for business, but think science provides only technological support.

CEOs do not speak to the ideological function of higher education in their public remarks, perhaps because they do not speak to higher education with any frequency. However, the Forum documents see higher education as a key ideological apparatus, although not a state apparatus. At this historical juncture, the Forum sees one of its major tasks as re-establishing the ideological primacy of the private sector, and sees higher education as critical to creating a climate that supports this goal. Great emphasis is placed on this task. Since this is not a view supported by the presidents when they speak independently, it is possible to attribute this point to CEOs and see it as a point of difference between presidents and corporate leaders.

The differences between the groups are probably rooted in the divergent ideologies used to legitimate the two sectors. The notion of the state as external to the university is in keeping with higher education ideology, which devotes a great deal of energy to fending off government intervention in the postsecondary sector, probably to avoid the constraints and low status that are attached to many state agencies in the United States. So too, the notion of advanced education and science as value-free, untainted by ideology, was devised and is routinely used to keep the university above the fray of politics. The "necessary fiction of the neutral university" eases the strain of research universities dependency on political allocations for federal science dollars, to say nothing of state infrastructure monies.

The Business–Higher Education Forum appears to be dominated by business interests. As the ratios of business to higher education concerns indicate, only in one report, *Corporate and Campus Cooperation*, are the concerns of each sector given equal weight. When there is mutuality of interest, as in *Corporate and Campus Cooperation*, it is not research partnerships for high technology enterprises that are stressed, but traditional relationships in which corporations offer a variety of forms of support in return for which universities provide training and legitimation of corporate interest, including multinational economic development strategies.

If business concerns are dominant, why are universities Forum participants? In the long term, universities have at least two strong reasons to work closely with the corporate sector. The reasons are not complicated or subtle. First, corporate leaders and university presidents share a set of institutional values. Second, universities have a history of aligning themselves with progressive leaders of large corporations who are likely to provide them with reliable resources, whether material, political, or symbolic.

The insitutional values shared by CEOs and university presidents are articulated throughout the series of reports. Each see themselves as part of the "private sector," even though some of the colleges and universities making this claim are indisputably public. The private sector is seen as embodying excellence, quality, innovation, independence, creativity, diversity, vision, and active agency, whether in corporations or on the campus. These qualities allow the institutions in question to claim special prerogatives: managerial discretion, autonomy, a voice in policy making, and almost unlimited social and economic resources. By declaring these values their own, universities become something more than state agencies subject to the heavy hand of public bureaucracies and the unpredictable forces of popular political currents.

Universities have historically aligned themselves with the liberal wing of the corporate sector, seeing these CEOs as predictable suppliers of resources and as reliable political allies.[16] The only period in the hundred-odd-year history of research universities in which they distanced themselves from corporations was during the years 1945–1965, when it appeared that the federal government would be more predictable and reliable and less demanding. When federal monies were no longer forthcoming at the rate to which the research establishment had become accustomed, it turned once again to the business community. The rekindled alliance seems to depend less on direct corporate support and more on corporations as political allies that are able to pressure federal and state governments for more funds.

In return for such support, universities legitimate the global development strategy which the multinational corporations participating in the Forum are developing. Although this exchange is pragmatic, it is neither cynical nor grudging. As corporate leaders want to regain the economic dominance over global markets that came with victory in World War II, so universities would like to reassert their intellectual hegemony in world scientific communities. And in many respects, research institutions benefit when multinationals do. Their presidents often sit on multinational boards. Universities are frequently stockholders in these corporations, and sometimes hold patents these corporations exploit, the international marketing of which will bring them increased revenues. Research institutions train the small cadres of highly specialized professionals who work in these corporations as researchers and managers.

Despite the lack of attention in the Forum reports to university development of entrepreneurial science, corporation and campus seem to share the same conservative ideology which stands in the way of the global development strategy they are promoting. For American multinationals to capture global markets in the manner suggested by the reports, public monies would have to be channeled to commercial research and development, world trade would have to become freer, and national planning mechanisms for penetrating international markets would have to be developed. Even as the reports struggle toward these goals, they turn away from them, unable to give up the mechanisms they recognize as problematic—the Cold War, the national security state, the primacy of the private sector in policy making—because they have proved so useful in the past in allowing corporations and universities to acquire unobtrusively the resources they have needed from the state. Yet until funds for military research and development are curtailed, monies for commercial development are unlikely to be available; unless trade patterns that include Soviet bloc countries are developed, global markets will remain elusive; and unless national planning agencies with some scope and power are developed, the coordination and promotion of successful market strategies is unlikely.

Despite the promises presented by formal partnership in a high technology era, the Business–Higher Education Forum is a rather conventional alliance. Although research and development partnerships for high technology and its application are important, emphasis seems to continue to be placed most heavily on the routine demands of science-based industry, training, and legitimation. The corporate community offers political, material, and symbolic support, in return for which the university provides training and legitimation for a multinational strategy for global economic development. Leaders from both sets of institutions stress their private sector status and the special privileges that stem from it. Both look to leverage more resources from the state through their alliance, yet are reluctant to abandon those mechanisms that have served them well in the past.

Conclusion

Having established the basis of cooperation and shared consciousness between CEOs and university presidents, I will try to put the Forum reports in the context of the several theories reviewed at the beginning of the preceding chapter. At issue is how to interpret the Forum leaders' policy positions and their posture toward the state. The ideology promulgated by the Forum, and the way it figures in the leaders' posture toward the state, is also important in assessing which theory best accounts for the relationships and joint policy efforts revealed by the reports.

Neopluralist theory does not seem to explain the Forum leaders' efforts. Business is clearly dominant, not partially or occasionally dominant. Indeed, business has expanded the scope of its interests to include almost all sectors other than labor. The university sector as well as the corporate sector advocate decision-making strategies that insulate them from the electoral process. Rather than seeking to influence the state via the electoral process, the Forum expects the bureaucratic state to organize the competitive energies of corporations to win a larger share of global markets. Moreover, Forum leaders expect the private sector to have a powerful voice in directing the agencies created by the state. This blurring of lines between corporate and state policy making does not fit well with neopluralism, which usually draws distinctions between private and public sectors.

Corporatism comes much closer to accounting for the policy developed in the Forum and its posture toward the state. The Forum sees the state as engaged in a multitude of functions at a number of levels, and would like the state to streamline and arrange its organizational structure hierarchically, putting the concerns of business at the apex. So long as business concerns are recognized as paramount, the Forum emphasizes cooperation and consensus among industry, education, and government.

However, the Forum does not systematically include labor as a sector worthy of representation in sectorial policy making. Although labor is occasionally mentioned, it is a group that must accede to the positions developed in the Forum, not participate in making them. Lower levels of unionization are noted with approval, as is unions' acceptance of lower wages. Many of the Forum policies, if realized, would further reduce traditional blue-collar jobs. The Forum economic development strategy is likely to move high-waged labor to less costly sites overseas, and to concentrate on high technology in the United States, leaving the service sector unorganized and underpaid.

The Forum's position on labor and its readiness to use the state to implement its policy are perhaps more closely captured by a neo-Marxian interpretation. The state is not sharply separated from civil society, and is "an expression, or condensation, of social-class relations, and these relations imply domination of one group by another."[17] Domination usually leads labor to struggle with the corporate class for more favorable treatment which is usually mediated and codified by the state through vehicles such as labor law. Although the Forum does not deal with labor as an organized entity trying to raise its voice in the policy arena with positions that counter its own, that does not mean that labor is not so engaged. Indeed, ignoring organized labor may be a means by which the Forum downplays the strength of organized labor and the possibility of class struggle.

The Forum's deep concern with ideology also seems to point to a neo-Marxian interpretation. The Forum is concerned with creating a "consensus" that brings the entire nation to subscribe to the importance of the private sector in international competition. One of the primary purposes of the Forum is to begin spelling out the ground rules for such a consensus. The Forum, then, in its own voice, although not using the exact words, sees the creation of ideological hegemony as one of its main tasks.

There are a number of other points that a neo-Marxian perspective does not address with great clarity. As formulated, the class struggle perspective is macrotheoretical and overly abstract, given to terms such as *"the dominant class," "mediation,"* and *"ideology."* These are not helpful when considering actual institutions such as the Forum. In the case of *"dominant class,"* the term is so broadly inclusive the principles of class membership and solidarity are often uncertain. In the case of *"mediation,"* the term is used as a broad descriptor for class and state interrelationships which benefit capital, obscuring the actual patterning of interaction. When looking at ideology, the neo-Marxian perspective usually concentrates on the dominant ideology, occasionally looks at opposition to it, but rarely examines competing ideologies within a class or ideologies developed by the middle class. For the most part, the neo-Marxian perspective overlooks the middle class, creating a serious gap in our understanding of concepts such as ideology and mediation, which usually turn on the middle class for implementation. In the following chapter, I will try to give the neo-Marxian perspective some grounding in the milieu in which the Forum operates.

CHAPTER 8

Class, State, Ideology, and Higher Education

In this final chapter, I will try to look at the implications of neo-Marxian theory for the Business-Higher Education Forum and postsecondary education. I will try to speak concretely to the data and arguments presented previously. The central concepts with which I will be concerned are class, state, and ideology, and how these play out in the postsecondary policy formation process. I will also be concerned with what these concepts say about how power is wielded. When the theory does not fit, I will try not to force it and will consider instead alternative conceptualizations.

Class

Inquiry into the social-class background of the Forum leaders indicated that they were neither rich and powerful nor poor and downtrodden. The majority of leaders of corporations and campuses were drawn from the small towns and provincial cities of America. There is little indication that any were from the working class, or conversely, that many were from upper-class families or established elites. The prestige of the institutions from which they received B.A.'s was quite mixed, but as the leaders moved up the educational ladder, taking higher degrees, they tended to be increasingly concentrated at elite, private research universities.[1]

The data on Forum leaders do not suggest that they are members of an upper class or owners of the means of production. Although corporate leaders in the Forum sample may have very high salaries and hold stock in the companies which they head, they do not seem to be part of that relatively small group, approximately 0.5 to 1.0 percent of the population, which holds about 25 to 30 percent of the total national wealth and has cohesive social-class connections.[2] University leaders, although highly paid by most standards, have less income and fewer stock options than the corporate leaders.

Although the Forum members may not be members of the upper class, they are heads of elite institutions. Of Forum CEOs, approximately 90 percent are in charge of companies in the top *Fortune* 200. Of university presidents, 56 percent are heads of AAU universities, and another 20 percent are heads of selective colleges. Approximately 75 percent, then, lead elite colleges and universities.[3]

If the Forum leaders are not members of an upper class, how can we think about them? Can they be conceived of as an interest association which works to influence policy, or part of a relatively unified power elite through which an upper class rules,[4] or a professional-managerial class with interests of its own?[5] I want to argue against the notion that the Forum is an interest association representing a specific sector. The Forum was constituted to bridge the gap between corporations and higher education rather than to represent a single sector. Moreover, the analysis of the various groups in which Forum members routinely participate indicated that they are strongly represented across a spectrum of sectors—corporations, education, foundations, policy formation organizations, civic and charitable organizations, and the military. The Forum seems to be one of a network of policy organizations in which CEOs and university presidents are active.

This network of policy organizations has discernable metes and bounds. There is little evidence that the group is connected to organizations or associations dominated by members of an upper class. Nor do Forum members regularly participate in organizations that represent blue- or white-collar workers, or ethnic or racial groups.[6] Forum leaders seem to have interests that cross sectors horizontally, but not vertically.

What, then, is the principle of horizontal organization? Are Forum members part of a managerial class whose interests are distinct from owners? If so, according to proponents of the "managerial revolution," their separate identity should be manifested by a broad and genuine concern for the public welfare. Forum documents do not convey such a concern; quite the contrary, they repetitively insist on the primacy of the private sector, even at the expense of higher education. The Forum notes the following in its second report: "Indeed, the need for increased investment in education must be balanced against the need for increased investment in productive capacity. No significant overall rise in corporate support of higher education can be expected until there has been a material improvement in corporate earnings."[7]

Rather than representing leaders who are products of a managerial revolution, concerned with the public as well as the corporate good, Forum leaders appear to operate on "classwide principles" similar to those articulated by Useem:

> In this framework, location is primarily determined by position in a set of interrelated, quasi-autonomous net-works encompassing virtually all large corporations. Acquaintanceship circles, interlocking directorates, webs of interfirm ownership, and major business associations are among the central strands of these networks. Entry into the transcorporate networks is contingent on successfully reaching the executive suite of some large company and it is further facilitated by old school ties and kindred signs of a proper breeding. But corporate credentials and upper-class origins are here subordinated to a distinct logic of classwide organization ... The unplanned consequence [of these networks] ... is the formation of a communications network that defines an inner circle of the business community in each country that can rise above the competitive atomization of the many corporations that constitute its base and concern itself with the broader issues affecting the entire large-firm community.[8]

However, I see these classwide principles as extending beyond the corporate sector to the academic and perhaps other sectors—public interest (mass media, foundations, law, civic and cultural organizations) and government (especially the executive and military).[9] In this expanded definition, classwide principles of organization depend not exclusively on ownership of the means of production, nor on wealth, but on a relationship to central social institutions in the public and private sector. These institutions, if formally defined, would probably be similar to Dye's operational definition of major societal institutions. This group of institutions is composed of businesses which control over half of the nation's total corporate assets, the several mass media corporations, the large prestigious New York and Washington law firms, the well-endowed private universities, the major philanthropic foundations, and the most influential civic and cultural organizations.[10] I would also include major public institutions, ranging from prestigious research universities to relatively insulated state agencies, such as the Supreme Court or the Federal Reserve, a point which will be further discussed in the section of this chapter that treats the state. Within this transinstitutional network, the leaders of the corporate sector are probably *primus inter pares* because of their command of resources and their relatively cohesive political goals, but the heads of other institutions are fully fledged class members when they engage in policy formation within the network.

Although I am arguing for the importance of central societal institutions, I am not taking an elitist position.[11] I see the interrelated, quasi-autonomous networks in which class-conscious institutional leaders operate as empowering them as policy actors, but not as automatically conferring power. The social institutions which they head are at one and the same time a power base and resources which they strive to control. In other words, some public institutions have become part of the means of production and are objects of contention in class struggle. Power resides in an institutional *class*, which encompasses actors in both the private and public sector, rather than in institutions themselves. Because power resides in the class and not the institution, the institutions themselves can still be the site of class struggle, and are open to control by various classes.

If Forum members are operating on classwide principles of organization that extend beyond the corporate sector, there should be indications in their reports of a shared consciousness that centers on an awareness and promotion of mechanisms for maintaining class welfare, power, and privilege. The Forum gives a great deal of attention to creating mechanisms for mediation which will serve the interests of the institutional class. Generally, the Forum inveighs against governmental "ad hocism," "protectionism" by labor, and firms which urge "special" needs, and instead calls for a broader vision of society in which central groups work together:

> American society, through its seeming addiction to adversarial relationships, has created a formidable barrier to restoring the nation's competitiveness. Yet any effort to rebuild U.S. economic competitiveness clearly requires consensus on the nature of the problem and cooperative action among business, government, labor and education to solve it. The nation has had few positive experiences with such cooperative efforts—except during times of war. Moreover, very few institutions are capable of facilitating such cooperation. This major gap must be overcome.[12]

The way to overcome the "gap" is through the state:

> America's national effort must allow each sector to perform the specific parts of the overall effort for which it is best suited and to promote cooperative actions when necessary. Thus, government's responsibility is not to direct the activities of the private sector, but to streamline its own processes and create an environment in which the individual and collective talents of the private sector can be focused to meet the competitive challenge.[13]

As the above passage indicates, the task of the state is not to mediate impartially between groups, but to foster the interests of key private sector institutions, which, as I will suggest in the next section, are important to the institutional class.

Specifically, the Forum tries to maintain class privilege and power through policy proposals, most of which are aimed at reinforcing the decision-making power of the institutional class. As we have seen, the legislative agenda of the Forum focuses on (1) revision of laws which restrict managerial prerogatives, such as antitrust laws and cross-industry regulation; (2) expansion of laws which increase private-sector control, such as tax code exemptions and intellectual property statutes; and (3) the strengthening or creation of state agencies to coordinate and plan for private-sector well-being, such as the proposed National Commission on Industrial Competitiveness and the Information Center on National Competitiveness.[14] The Forum, then, does promote policy which develops mechanisms that mediate class relations in ways designed to privilege its members.

Although Forum leaders agree on promoting policy that keeps decision making in the hands of the institutional class, they sometimes seem indecisive about the exact shape these mechanisms should take. They reveal the highest degree of ambivalence toward the military, the Cold War, and national planning. They are critical of using the military and the Cold War to subsidize business, to develop research and development, and to expand the national security state. Their criticisms are based mainly on the wastefulness that accompanies what is largely indirect subsidy. But historically they have given military leaders access to the institutional class and have benefited greatly from defense funding that has unobtrusively provided them with extensive federal monies.[15] Although the Forum wants more open and direct coordination of nonmilitary economic policy at the national level, it hesitates to offer concrete mechanisms that are open to public scrutiny and that could be taken over by members of other classes.

Forum leaders, whether CEOs or university presidents, seem committed to developing mechanisms for mediation which enhance the power of the institutional class. However, in some areas of joint interest there is not a neat fit between CEOs' apprehension of corporate needs and university presidents' understanding of what is good for research institutions. Tension between the two groups over issues in these areas could possibly lead to conflict.

At first glance, the issue that seems most likely to produce conflict is the free flow of information. University leaders insisted on this freedom, which did not seem strongly supported by corporate leaders and could be construed as contrary to corporate interests in ownership of knowledge. However, university presidents are not unreservedly committed to the free flow of knowledge. Like corporate leaders they want to enhance their claims to intellectual property, a position which hinders the unfettered movement of knowledge. This shared position lessens the possibility of any enduring conflict between CEOs and presidents. Moreover, corporate leaders are committed to constraining state regulation and preserving free markets, positions which would presumably lead them to support university presidents efforts to maintain a free flow of knowledge, so long as it is not likely to impede ownership. In sum, both sets of leaders are likely to struggle jointly and vigorously against state regulation, as in the case of military classification, especially when property rights are at stake or when their prerogatives as high technology managers are threatened.

The area of greatest potential conflict may stem from the divergent interests of the two groups. Corporate leaders are focused on increasing America's economic competitiveness and see a major mechanism for accomplishing this as re-establishing the power and privilege of the private sector. University presidents are deeply committed to increasing the prestige of American universities and see as a major mechanism the continued flow of federal and state dollars for research and for general appropriations to higher education. The CEOs want to constrict the flow of state monies to any other than corporate enterprises while university presidents want to channel substantial state resources to research and higher education. These differences are mitigated by their joint emphasis on business and higher education partnerships, which label the flow of state resources to higher education as monies for strengthening corporate ventures, but nonetheless the differences remain.

The question arises as to what the university presidents would do if concerns specific to their institutional sector clashed with interests expressed by corporate leaders. According to my analysis, the allegiance of these leaders is to the institutional class rather than to a specific sector. The university leaders' concern with classwide prerogatives would probably outweigh their commitment to higher education as a whole, or more exactly to all of postsecondary education. If forced to choose between the sector as a whole and their commitment to Forum goals, the presidents would probably choose the later. They would make such a choice because they understand their elite research institutions, whether private or public, to be part of a wider transinstitutional network of privilege which must be sustained as a whole.

That AAU presidents might be faced with such a choice in the near future is quite probable. Given the federal deficit, the several states that require annually balanced budgets and the ever expanding need of the postsecondary sector, there does not seem to be enough public money to go around. Rather than arguing the claims of students generally or defending institutions of higher education as whole, AAU presidents are likely to support more monies for research and excellence in graduate education as the most critical needs in the postsecondary sector. Indeed, their Congressional testimony indicates that this is the position they have historically taken. Although there are merits to this position, it is certainly not the only or even the obvious position that leaders of postsecondary education might take to improve the flow of resources. Public and legislative interest at both the state and national level are currently turned to undergraduate education, not graduate education or research. It is easy to see how leaders of four year institutions could make a strong case that quality undergraduate education is the most efficient and economical way to lay the ground work to meet personpower needs for an internationally competitive, high technology economy.

However, given AAU presidents' allegiance to the institutional class and their concern with maintaining their relations to the transinstitutional network it controls, they are unlikely to support popular arguments for funding and reform of higher education. Non-elite institutions, whether graduate schools or four year institutions, fall outside the rather tight circumference of the circle of privilege. Expanding the circle would threaten the cohesion, position and entitlements claimed by the institutional class.

I have argued that the leaders of the most prestigious research institutions would put the interests of the institutional class before those of the postsecondary sector as a whole. I see this as a relatively unacknowledged policy problem facing the postsecondary sector, one that we fail to examine closely for several reasons. First, most of the policy scholars in higher education work at research universities and are committed to the logic that places them at the apex of the higher education system. Second, we usually treat science policy and graduate education separately from undergraduate education, and therefore do not see clearly the inherent conflicts between the two over resources and educational priorities.

I have answered the question of what AAU presidents would do if forced to make a choice between the institutional class and the postsecondary community as a whole by suggesting that AAU members are part of the institutional class and would naturally support it. However, my answer to this question flows from my analysis of the class position of Forum leaders and should be treated skeptically. Questions about the points at which the presidents of elite research institutions might break solidarity with CEOs can and should be answered empirically, and will provide an indication of whether or not the concept of an institutional class is a useful one.

Just as the relation of presidents of elite research institutions to CEOs is an important question for understanding the class position of leaders of higher education, so is their relation to faculty. According to my definition of class, AAU university presidents are part of an institutional class which tries to operate as a functional ruling class. As such, do they represent their faculties, or are they at odds with them? Although we have not considered the position of faculty, in order to work out the implications of the institutional class for postsecondary education, we need to try to think through what the class position of faculty is and how it articulates with that of the institutional class.

On many points, I agree with the Ehrenreichs' definition of the Professional-Managerial Class (PMC), and see faculty as belonging to it. According to the Ehrenreichs, "the Professional-Managerial Class ... [consists] of salaried mental workers ... who do not own the means of production and whose major function in the social division of labor may be described broadly as the reproduction of capitalist culture and capitalist class relations.[16] For the Ehrenreichs, "the university is the historical reproductive apparatus of the PMC and a historic center for the production of new knowledge, disciplines, techniques, heresies, etc.: both functions which have acquired a semblance of autonomy." They see the class as "derivative," its existence presupposing a social surplus. Although the PMC is nonproductive, it is nevertheless essential because "the relationship between the bourgeoisie and the proletariat has developed to the point that a class specializing in the reproduction of capitalist class relationships becomes a necessity to the capitalist class."[17] The function of the PMC is to service mass institutions of social control aimed at promoting working-class acceptance of their place in the social structure. Although the PMC deals regularly with the working and under class through social-service agencies and schools, relationships between the PMC and these classes are inherently antagonistic. Similarly, the PMC depends on the bourgeoisie, but has always possessed "a class outlook distinct from, *and often antagonistic to*, that of the capitalist class"[11] (italics in original).[18] The PMC is between labor and capital, but has some autonomy of its own.

Unlike the Ehrenreichs, I do not make the conventional neo-Marxist differentiation between productive and non-productive labor, a differentiation which labels most who work primarily with their minds as non-productive. Nor do I see the state as a non-productive sector. For example, I see faculty as engaged in more than the reproduction of culture and class relations; I see them as providing training and new knowledge without which capitalist production cannot occur. Faculty have a technical as well as an ideological function in the social division of labor and the technical cannot be ignored. As indicated earlier, I think that some state institutions have become part of the means of production and that those who work at managerial tasks in these institutions are producers.

The distinctions I have made allow analysis of the faculty as other than a collectivity. There are faculty and there are faculty. Some work at universitites which are part of the transinstitutional network over that the institutional class tries to maintain control, others do not. There are marked differentiations between faculty even at the elite research institutions that are part of the transinstitutional network. Some have extremely sophisticated technical skills which they turn to production of high technology; others do not. These differences strongly affect faculty relationships to the institutional class.

Although there are obvious differences between faculty, to some degree they have a shared consciousness and resource bases that are distinct from the institutional class that heads the organizations in which they work. Their class consciousness does not turn on preserving managerial prerogatives, maintaining hierarchy, and promoting an ideology which puts economic prosperity first. Instead, faculty have developed a "professional" consciousness that hinges on concepts such as specialization, professional autonomy, collegiality, and an ideology of expertise, which are at odds with managerial authority. Although faculty are dependent on their institutions, and therefore the institutional class, they also have a resource and power base of their own which is distinct from their institutions. Most faculty are members of professional or learned associations which represent their disciplines. Very often they are members of transprofessional organizations, such as the AAUP, the AFT, or the NEA. Often they are able to provide services to clients or engage in consulting, generating individual income independent of their institutional position.

Although faculty have a distinct class consciousness and an independent resource and power base, they are highly fragmented. Faculty are probably least autonomous but most powerful the closer they are to the institutional class. At first glance, this appears to be counterintuitive, especially with regard to autonomy, since faculty at prestigious institutions are reputed to have more autonomy than any others.[19] Faculty at prestigious institutions do have a great deal of occupational autonomy, if this is taken to mean the ability to order their own time and to set goals without managerial direction. But these faculty are also much more likely to be involved with the institutional class through large research and development enterprises, entrepreneurial relations with corporations and state agencies, or consultancies with these organizations. In this capacity, their power to participate in and influence the decision-making process is high, but the likelihood of their developing policy at odds with the institutional class is low. This is especially likely to be the case in the professional schools, which now account for about half of all university faculty, and in areas in the sciences with entrepreneurial potential. For example, faculty in the biotechnology departments at Harvard and the University of California are probably virtually indistinguishable from members of the institutional class as represented by Forum leaders.[20] Faculty at elite institutions, then, have a great deal of occupational autonomy but may lack authentic autonomy.

However, there are areas within prestigious universities and colleges that are part of the transinstitutional network over which the institutional class tries to maintain control where faculty can be quite distinct from the institutional class. These areas are located in certain fields of the social sciences and the humanities as well as in professional schools that do not connect closely to organizations critical to the institutional class, areas such as education, social work, and nursing. In these areas, it is possible for faculty to develop a more authentic autonomy, but there is less likelihood that they will influence policy and decision making.

Faculty in undistinguished four-year institutions and community colleges probably have greater professional space than faculty at elite universities to develop positions distinct from the institutional class across a variety of issues, which could range from modes of occupational organization to demands for the expansion of state services. At the same time, faculty at these institutions have less power and influence because of their distance from the institutional class. To become effective in policy arenas, they would have to develop alternative political networks and find ways to empower them, as well as expand their economic base, all daunting tasks. Moreover, faculty in undistinguished postsecondary institutions are influenced by faculty at prestigious schools to accept the primacy of the institutional class in decision making. Finally, these faculty are divided by their access to rewards regulated by the institutional class—for example, salaries, status, and consultancies. There may be more space at these institutions for an autonomous faculty to occupy, but there are also many opportunities for fragmentation.

The end of rapid expansion of the postsecondary sector in many state systems means that there is an intensified struggle for increasingly scarce resources. The part that faculty will play in this struggle may be determined by their relations to the institutional class. As we saw in Chapter 5, the traditional response of leaders of prestigious institutions, usually with the full support of their faculty, has been to promote stratification which insulates them from the rest of the postsecondary sector. The increasing distance between faculty in the professional schools and entrepreneurial science and other faculty, and between faculty at elite research institutions and other institutions, in terms of salary, prestige, and occupational autonomy, may create conditions that allow some faculty to develop a voice different from that of the institutional class. These faculty might turn away from the institutional class and toward other classes and representatives of social movements. What the consequences of this alignment would be for postsecondary education is unclear. The connections between the various postsecondary sectors, the class structure, the polity, and the economy are undertheorized and need careful consideration by scholars of higher education.

The State

Many recent theoretical debates have turned on the question of "the relative autonomy" of the state.[21] These inquiries focus on whether or not the bourgeoisie, with its control over the economy, dominates the state, and if this is not the case, under what conditions the state operates independently. The question as posed is usually too abstract, too macrotheoretical, and too often an acid test of the purity of scholars' commitment to a neo-Marxian theoretical perspective.[22] A great many theoretical assumptions are contained in the question, and these have very often not been addressed. I would like to reformulate the question in a way that reexamines some of these assumptions, and consider higher education in general and the Forum in particular, as a case that casts light on the connection between class and state.

I think the question of the relation of civil society, or the private sector, to the state is posed in a way that is falsely dichotomous. The two are always treated in opposition to each other, as clearly demarcated and definatively different. What if we think of civil society and the state as so permeating each other in the post–World War II era that it is no longer analytically useful to see them as separate? One way to see whether or not this line of inquiry has any value is to ask if the state and the private sector can be conclusively defined as different in terms of the characteristics usually assigned to one sector or the other. If the characteristics of private enterprise are seen as private control, capital accumulation, profit making, and competition, can we make the case that state agencies do not engage in similar behavior? Let us take higher education as a case in point, looking to see if private higher education can be separated from public, and both from private-sector economic institutions with regard to control, resource base, organizational goals, and conditions of operation.

Although private higher education has a formal legal status that allows it to choose self-perpetuating boards, there is little else that separates it from public higher education in terms of control, especially at the level of prestigious research institutions. The boards of private and public institutions even tend to have roughly the same social composition,[23] varying with the type of institution. In terms of oversight, both are subject to a variety of forms of monitoring by agencies in private and public sectors: alumni, state boards, federal regulatory and civil rights agencies, and private organizations such as accrediting and professional associations. In terms of resources, private higher education institutions frequently receive from a quarter to half of their annual operating expenses from the state, at its several levels, and elite public institutions are the recipients of substantial private gifts.[24] In terms of staffing, faculty at prestigious institutions in both the private and public sector tend to have a strong, usually decisive, voice in hiring and firing. Private education, then, may be difficult to distinguish from public, especially at research institutions, if criteria such as oversight, resources, and personnel are used.

Distinctions between higher education, whether private or public, and other private sector organizations, such as corporations, are not as hard and fast as they initially appear. For example, many higher education institutions are now engaged in profit-making enterprises. Their activities run a broad gamut and include exploitation of patents through direct development as well as leases and royalties, joint ventures with faculty in high technology businesses, incubators, research parks, and university-industry-state partnerships. Indeed, many AAU institutions now have for-profit corporations. As the examination of Forum documents indicates, there is a strong effort to recast the university's public service function as economic development, which includes an array of profit options for higher education.[25]

In contrast to the profit-making activities of some public and private universities, postsecondary institutions habitually categorized as part of the private sector are not often profit oriented. Selective independent colleges, for example, are uniformly included in the private sector but are not concerned with profit making. Foundations play an important role in postsecondary education, are routinely defined as private, yet are not concerned with accumulating profit, and in fact have to disperse profits at regularly scheduled intervals.

Public and private higher education institutions often engage in competitive behavior not dissimilar from that of private corporations. Postsecondary institutions compete for students, faculty, research monies, and gifts. They work as partners with business in highly competitive economic development activities, ranging from supercolliders to polymer production. At the same time, there are private sector companies which do not engage heavily in competition. Most of these are companies which, for a variety of reasons, have been granted public monopoly status, such as public utilities and public broadcasting companies.

Even formal differences between the private and public status of organizations and entities are fluid. During the expansion of postsecondary education in the 1960s, private institutions sometimes became public, as was the case with the University of Buffalo, which was purchased by the state and turned into SUNY at Buffalo. So too, in the economic sector, enterprises are sometimes transferred from the private sector to the public—for example, urban transit systems, banks, grain elevators, and railroads, usually when they are badly managed or no longer profitable.[26] Conversely, under the Reagan administration, there has been a push for "privatization," which involves selling government assets or enterprises to the private sector and contracting public services, such as prisons, to profit-making organizations.

Although many organizations in civil society may fall neatly into the category we usually label "private," and many state organizations into the one we call "public," these definitions are not very firm or precise, especially at the level of central societal institutions. Like so many other categories, they are socially constructed. Rather than thinking of private and public as meaningful dichotomies, it might be more fruitful to think of private as a label that systematically privileges those institutions so named, regardless of their organizational control, goals, functions, and performance.[27] If institutions are designated as part of the private sector, or not part of the public—as is the case with institutions such as the Federal Reserve, the Supreme Court, and the National Security Council—they are usually able to command more resources, greater managerial prerogatives, and more autonomy. This label allows resources to be concentrated on central societal institutions, which are then available to the institutional class.

An instance of the social construction of the private and public sector is seen in the Business–Higher Education Forum's systematic designation of all of higher education as part of the private sector. The Forum does this continually.

> Therefore, the private sector—industry, higher education and labor—should strive to exploit and should be encouraged to exploit this new technology.[28]

> While the Forum recognizes the important role played by the federal government in developing the U.S. space program, it encourages more involvement by the private sector, specifically business and higher education, in future space endeavors.[29]

In defining all of higher education as part of the private sector, the Forum is aligning higher education with the institutional class and attempting to privilege the postsecondary sector, especially the prestigious institutions that constitute the majority of the Forum membership.

If civil society and the state are not definitively separate, then the state, like civil society, becomes a site of class struggle.[30] Some state sectors are dominated by the institutional class, and others are relatively "insulated"; some are contested, and still others are dominated by the PMC or the working class.[31] Although central societal institutions tend to be in the hands of the institutional class, there is always struggle over who controls them, and, over time, control can change hands, in part or altogether. With regard to the state, the outstanding question is what principles order the degree of insulation, or even autonomy, of various sectors, branches, and levels of institutions which are commonly assigned to the "public" category. I would like to consider briefly higher education as a relatively insulated state institution, using as a working hypothesis the idea that the more central a university is to maintaining a flow of resources—broadly construed as money, power, privilege, and ideology—to the institutional class, the less insulation, and the less central, the more insulation.

Prior to World War II, prestigious public universities were relatively insulated state agencies.[32] Sometimes insulation was formal, as was the case with those public universities whose autonomy was guaranteed through state constitutions; more often, it was informal. Informal insulation or (always relative) autonomy was generally established and maintained by an emerging PMC interested in carving out a base for itself independent of capital or labor. The PMC constructed a series of organizations—professional associations, learned associations, the National Research Council, the Social Science Research Council, the American Council of Learned Societies, the American Council on Education—that helped them negotiate between civil society and the state, to claim some autonomy when they operated in either.[33] Although universities were probably more dependent on the corporate sector than not in the period prior to World War II, they were able to distance themselves. As scholars of higher education have noted, this distance was achieved at least in part because universities were not yet embroiled in social processes such as stratification, national security, and high technology development.[34]

During World War II, university science emerged as important to national security. This made science more central, but ironically, less autonomous. The flow of federal dollars to research universities increased enormously, initially through the DOD and DOE and, after

1963, through the National Institutes of Health.[35] As we saw in Chapter 4, these monies, although aimed at quite specific goals of the mission agencies, were usually described by a rhetoric of "pure science" and "basic research." This language allowed universities to preserve the distance between themselves, the federal government, and the corporate sector even as they came closer together in terms of dollars and joint projects. With regard to DOD and DOE, the secrecy invoked by national security, together with the technical expertise required for high-energy physics, further shielded the university from encroachment from state and corporate oversight, even as the university grew more dependent on defense dollars. At the same time, university participation in national security served to legitimate research and development for corporations and to shield military dollars from careful accountability. Essentially, the institutional class—in this case, prestigious research institutions, the military, and corporations engaged in executing defense contracts—was able to channel public resources toward the organizations central to its welfare, even as close oversight and public scrutiny were avoided.

This relationship between university, corporations, and the state came under severe criticism during the Vietnam War. On the one hand, the state began to demand more concrete contributions to the war effort from the scientific establishment. On the other, students questioned the use of state and university resources for defense, and, indirectly, for corporate subsidy. Great conflict characterized university campuses, and much greater scrutiny was directed towards universities. Or, as science became more central to sustaining the flow of resources to the institutional class, it became less insulated, from both state pressures for performance and public pressure for accountability.

Currently, there are two major thrusts in terms of science policy, both of which are likely to increase the influence of science while decreasing its autonomy. First, there is the development of business-university-state partnerships for economic development. As high technology takes center stage, the differences between basic and applied research are collapsing. Since the mid-1970s, government support for university-industry partnerships has grown rapidly, with the National Science Foundation (NSF) funding an array of such programs.[36] At the state level these partnerships emphasize the importance of basic or fundamental research for the creation of high technology that will reinvigorate American industry and preserve its place in global markets. Although these partnerships increase the flow of resources to science, university research becomes less autonomous because of its centrality.

Representatives of corporations and state government often have a voice in determining what projects will be funded.[37] Formal agreements between corporation, university, and state usually govern credit for the discovery, assignment, and exploitation of patents, and the time period before which sponsored research can be reported in scientific journals.[38] There is increasing concern about how much the economic success of academic work will figure in promotion and tenure decisions.

Overall, business-university-state partnerships may be changing faculty's role. Formally, that role is defined as tripartite, encompassing teaching, research, and service. The research aspect of the role may be in the process of being redefined to focus on commercially viable research, and the service aspect to unleash the entrepreneurial energy of faculty. Although these redefinitions may result in creative new work and configurations of resources, they will also serve to align faculty at research universities with the institutional class.

Second, under the Reagan administration, a strong initiative to revitalize the military-university-industrial partnership characteristic of the 1950s and early 1960s has been initiated. DOD funding for universities grew dramatically after 1980, from $495 million to $930 million in 1985, an increase of 89 percent. The National Science Foundation, the second fastest growing source of support for university research, grew only 51 percent during the same period. If DOD funding of off-campus institutes, such as MIT's Lincoln Laboratories, is figured into the equation, the Pentagon is outspending NSF on academic research.[39] As noted in Chapter 2, this DOD initiative has created problems for the university in terms of the free flow of scientific information, classification, and the global movement of faculty and students. Again, as science becomes more central to the defense effort, the university becomes less autonomous.

Because the place of the military in the institutional class is unclear at present, university presidents and CEOs have worked to limit national security encroachments with regard to the development, flow, and publication of corporate and university research. However, the outcome of the negotiations between these groups is still uncertain, and the centrality of research to technological development means that the institutional class will have a voice in determining the outcome. Whether or not the outcome favors positions traditionally taken by the university will depend on the degree of accommodation the institutional class reaches with the military.

The enormous expansion of the university in the 1960s provided opportunities for groups other than the institutional class to capture resources. The proliferation of institutions and lines was an opportunity for the PMC as well as for groups previously excluded from the university to secure resources. Although the PMC and excluded groups were able to make inroads at most postsecondary institutions, in private as well as public sectors and at schools ranging from research universities to community colleges, state institutions other than research universities probably offered the most opportunity, since these schools were outside the flow of resources patterned by the institutional class.

The PMC made its claims through the proliferation of specialties, licensure, and certification, linking education to the control of careers through state authority. When these fields were not central to sustaining the flow of resources to the institutional class—for example, in areas such as school psychology, nursing, and social work—there seems to have been little effort to control such expansion, and the PMC was able to gain a degree of autonomy. However, there are limits to the ground the PMC can claim. Given the resource shortages of the 1970s and 1980s, the institutional class began to move to contain expansion of the PMC. The institutional class is not so much punishing the PMC as attempting to guarantee its own share of resources.

Groups outside both the institutional class and the PMC made gains as state higher education expanded. In response to political protest, demands put forward by popular social movements, and lobbying efforts, state lines as well as positions in private institutions were allocated to departments of black studies, women's studies, labor relations programs, peace studies, and the like. Neo-Marxists, critical theorists, and feminists began to gain a tenuous legitimacy in some fields. However, neither the institutional class nor the PMC are committed to sustaining these gains in times of austerity. Therefore, programs that embody the gains of previously excluded groups have difficulty protecting their institutional base. Although these programs have maintained a lively presence in many colleges and universities throughout the postsecondary sector, it remains to be seen if they are "institutionalized" to the point where their existence no longer depends on the political efforts of their external constituencies, as well as on the professional efforts on their faculty.

In sum, it might be profitable to begin thinking about the principles that order the degree of autonomy possessed by state agencies. Although higher education as a whole is important to the institutional class, as indicated by Forum efforts to redefine all of the postsecondary education as part of the private sector, it is apparent that some segments are more insulated from the institutional class than others. I have suggested that those segments of higher education not critical to maintaining the flow of resources to the institutional class are most insulated. However, as a state apparatus, higher education is also a site of class struggle. This struggle crosscuts postsecondary education and may engage insulated fields and institutions, especially in periods of social or economic crisis. For example, women's studies programs and critical social science departments may forfeit their insulated status if they come into overt conflict with the institutional class. Or community colleges which become hosts for social movements supporting populations previously excluded from postsecondary education may have their resources cut. Although these principles, upon a more extensive investigation, might not prove to order insulation or autonomy with regard to higher education, they nonetheless suggest that autonomy within state apparatuses is not a fixed characteristic but is socially constructed, and that the rules which govern the relations that produce autonomy are a central problematic for students of postsecondary education.

Ideology

One of the major contributions of neo-Marxian theory to thinking about schooling as a state function turns on ideology. Ideology is viewed as one of the major weapons of class struggle, the one that persuades the majority of the citizenry to accept existing power arrangements, despite their many inequities. Postsecondary education is one of many "ideological state apparatuses."[40] It is perhaps the most important of the ideological apparatuses because it generates knowledge used in schools and trains teachers.

A major question for theories of ideological reproduction is how knowledge acquires its class content. I would argue that organizations like the Forum mediate between the institution class and the PMC, creating a classwide ideology which articulates with professional ideologies of expertise. The Forum can be seen as an effort on the part of the institutional class to draw the university into its vision of ideological hegemony. The Forum "does not impose its own ideology upon the allied group; rather it represents a pedagogic and politically transformative process whereby the dominant class... articulates a hegemonic principle that brings together common elements drawn from the world views and interests of allied groups."[41] If this is the case, the Forum should clearly promote an ideology that serves the institutional class, is congenial to at least some fractions of the PMC, and uses the rhetoric of the common good.

The Forum clearly sees the creation of ideological hegemony as one of its central tasks. In its initial report, the importance of presenting a coherent set of ideas for interpreting national economic problems is stressed: "For their part, business, education and labor must become more dynamic and flexible. In the past, a large number of specific recommendations have been made, each designed to solve a specific problem. The missing link has been an agreed-upon framework for the many actions, in the public and private sectors, that will constitute the American response."[42] The importance of education in creating and maintaining this "agreed-upon framework" or ideology is well recognized by the corporate leaders participating in the Forum:

> Education is of immense importance in economic growth and development, in the determination of standards of living, and in the formulation of the political arrangements that govern the relationships between business and the rest of society. Indeed, the climate in which business operates is a product of the social attitude toward business in general and toward the conduct of corporate enterprise in particular... In the long run, the totality of ideas fostered by higher education is a major determinant of the kind of society desired and the means by which it is achieved and preserved. Business has an obvious stake in the outcome of this determination, and one of the channels though which it can influence that outcome is through support of colleges and universities.[43]

In the main, the Forum promotes the values of the institutional class through an ideology made palatable to all groups through the promise of prosperity. The private sector, broadly construed, is preeminent. The Forum exalts an entrepreneurial culture, constrained by a corporatist framework, in which capital is the dominant force. The private sector is presented as embodying excellence, quality, innovation, independence, creativity, diversity, vision, and active agency, whether in corporations or on the campus. These qualities are seen as essential to lead us toward economic recovery. To ensure their successful leadership, CEOs, university presidents, and the heads of central social institutions need broad powers for decision making: managerial authority unconstrained by the popular branches of the government, autonomy, a decisive voice in public policy making, and almost unlimited social and economic resources. The entrepreneurial ideology of the institutional class, then, privileges its members by presenting them as essential to the economic salvation of society.

Although the Forum discusses the advantages of its leadership as broadly benefiting the general public, not much attention is given to identifying diverse political interest groups or clear constituencies and then engaging them in "a pedagogic and politically transformative process... [that] articulates a hegemonic principle that brings common elements drawn from the world views and interests of allied groups."[44] Instead, the Forum seems to direct its ideological energies toward the PMC. Labor is occasionally mentioned, but no effort is made to include it in participation in the partnership between business and higher education. An underclass is alluded to, but as a social grouping to be feared, not incorporated. Indeed, the Forum plays to PMC fears to ensure its alliance with the institutional class. It strongly emphasizes what the PMC will lose in terms of its standard of living and way of life if global competitiveness is not maintained.

Instead of creating a world view that embraces society as a whole, the Forum's ideological energy seems to be directed toward the upper reaches of the PMC, located in the prosperous professions, managerial positions, and the university. The reports essentially outline the opportunities available to PMC members located in central social institutions. Implicitly, the reports offer this fraction of the PMC the same privileges that Forum leaders claim for themselves. Given the focus of Forum attention, its members probably see the PMC as the swing vote between the institutional class and the working and underclasses. The Forum wants to align the PMC with the institutional class through a shared entrepreneurial ideology.

Although the Forum is trying to craft an ideology that includes the PMC, an ideology that celebrates the entrepreneurial values of the institutional class is not necessarily congenial to the PMC. Indeed, the PMC has been at pains to create its own ideology, an ideology of expertise that insulates it from the institutional class. The ideology of expertise centers on the university as a site that prepares professionals and on the professions themselves. Its infrastructure consists of somewhat autonomous professions and a relatively independent state able to underwrite professional training and to fund faculty lines. The ideology of expertise promotes a view of knowledge that is "scientific," "objective," and "value-free." It confers authority on the PMC by making experts the obvious arbitrators between contending groups in society with regard to policy and planning issues.[45]

The values enshrined in this ideology are specialization, authentic professional autonomy, collegiality, and concern for the general welfare rather than special or self-interests. Although academic expertise always has both a technical and an ideological function, the credo of the expert does not celebrate the ideological. The ideological dimension, if recognized at all, has been downplayed because acknowledging it undermines scientific expertise as the discourse for public decision making. However, the PMC implicitly formulated its ideology of expertise in complementarity with the institutional class. During the postwar period, the academic fraction of the PMC participated in developing the Cold War consensus that informed the economic and national security strategies of the institutional class in exchange for the opportunity to utilize its expertise in policy making. For example, prestigious academics worked as experts in developing the Marshall Plan, the nuclear power and armament industries, and the AID enterprises. While engaged in these activities, the PMC celebrated "the end of ideology,"[46] blind to the ways in which their expertise legitimized existing structures of privilege. This lack of clarity with regard to the way in which the ideology of expertise articulated with the institutional class and the PMC's own interests was not the result of any duplicity, but stemmed from their sincere belief that prevailing power arrangements best served liberal democracy. So long as prosperity was widespread and the PMC relatively small, an ideology of expertise formed in complementarity with the institutional class was not called into question.

But ideologies of expertise can be shaped in complementarity with groups outside the institutional class. The expanded, heavily state-funded university provided in large part the resource base for the growth of fractions of the PMC that used knowledge in diverse ways. As the university expanded in the 1960s, professors began to engage in generating alternative forms of knowledge, exchanging their expertise with external groups dedicated to challenging the institutional class—civil rights groups, women's groups, antiwar and peace activists, antinuclear organizations, environmentalists, and the like. These external groups contributed symbolic support and some resources, and very often used dissident professors' expertise to fuel reform.

Simultaneously, dissident professors were able to capture state resources for alternative forms of knowledge by appealing to the PMC ideology of expertise. They too valued specialization, autonomy, and collegiality, and demonstrated a concern for the general welfare as opposed to special or self-interest. Indeed, this fraction of the PMC saw themselves as the true custodians of the ideology of expertise, since they were able to see its limitations and understand its legitimizing function. Using the language of specialization, scholarship, and expertise, dissident professors were often able to secure funds for the expansion of new specialities and fields. These new areas were sustained by state lines and university resources. Among the university resources were new departments and courses, journals and library subscriptions, and fellowships and sabbaticals. In a very real sense, the state provided the resources to support academics critical of the institutional class. Dissident professors were able to construct their own academic networks, exchanging ideas with one another and consuming their own scholarship. They were also able to make their presence felt in professional associations and learned societies, many of which developed radical and feminist caucuses in the late 1960s and 1970s.[47]

Professors developing expertise at odds with the institutional class usually offer ideological and technical alternatives to current policy. In the area of foreign policy, for example, professors have worked with a wide variety of external groups to curb U.S. interventionist policies in Central America, particularly in Nicaragua. In many cases, the groups with which they work ask for more than just an end to war; they present carefully thought out policy alternatives for the region. Thus, the institute for Policy Studies drew on a number of professors in writing *Changing Course: Blueprint for Peace in Central America and the Caribbean*.[48] A great many professors have endorsed this long-term development plan for the Caribbean Basin. Academics have also exchanged expertise with the Institute for Food and Development Policy in attempts to help Central American countries become self-sustaining in terms of food production. Human rights organizations such as Americas Watch and Amnesty International have professors contribute to the production of annual reports that chronicle the abuses of various regimes. The form of the reports follows that of scholarship, giving a great deal of attention to accuracy, reliability, and verification. Religious organizations such as Witness for Peace and Pledge of Resistance often draw on the skills of liberation theologians housed in universities for their work in Central American countries.

Central America is only a single instance. State-supported professors exchange expertise with a wide variety of external groups trying to change prevailing policy. Examples of such exchanges are seen in professorial work with critical intellectual centers such as Public Interest Research Groups, the Council on Economic Priorities, the Union for Radical Political Economics, with single-issue groups such as nuclear disarmament and pro-choice organizations, and with religious groups trying to create new theologies. The domestic policy alternatives generated by these organizations and associations often call for an expanded and more active state, one able to offer maintenance and opportunity for the citizenry as well as planning and regulation of the private sector economy. The foreign policy alternatives are usually anti-imperial and anti-interventionist, aimed at creating a world order and a world-economy that allows third world countries a measure of autonomy and prosperity.

Ideologies of expertise can be used by the PMC in several ways. They can be used in complementarity with the institutional class, to service the state, to affirm the position of diverse groups with a variety of resources but without any broad social purposes, or in complementarity with popular social movements. The PMC rarely develops ideologies of expertise in complementarity with extreme right wing groups, perhaps because of the right's disaffection with the state, which is so necessary to maintaining the privilege of professionals.

The rewards offered by the institutional class are usually the most immediate. When PMC members work with the institutional class, they have access to policy networks and a place in the corridors of power, and often experience a sense of autonomy in their capacity as advisors and experts. However, working with the institutional class usually means working within a framework the boundaries of which are quite fixed. Expertise, in its technical and ideological dimensions, has to fit within the frame. The Forum, for example, would probably not be able to tolerate expertise that constrained multinationals to any degree.

Many faculty use their expertise to serve mass institutions of social control aimed at promoting working or under class acceptance of their place in the social structure. Generally, these are state institutions, such as schools or social welfare organizations. The rewards offered by this sector are not as extravagant as those associated with the institutional class. Professors engaged in offering this kind of service are often able to make some real improvements in these institutions, to enhance the legitimacy of the ideology of expertise, and to augment their incomes. However, their influence is by and large confined to making improvements that are powerfully, although certainly not completely, constrained by the demands of the existing order.

The rewards offered by participation in popular social movements are less tangible. They center on mortal righteousness, a sense of citizen efficacy, and the creation of community. Few boundaries are imposed, giving professors a great deal of authentic autonomy. However, tangible resources and rewards, including state positions, are not routine, a condition which forces these professors to work constantly to maintain popular and professional support, creating another kind of dependence.

At least partly in response to professorial use of state resources to further an expertise quite critical of present policy, business leaders have started to put considerable money into foundations that support the creation of a more conservative expertise. These organizations began to prosper in the early 1970s: the American Enterprise Institute; the Heritage Foundation; the Hoover Institute on War, Revolution and Peace; the National Bureau of Economic Research; the Center for the Study of American Business; the American Council on Capital Formation; and the Institute for Educational Affairs. These foundations have sometimes received very large donations from businessmen known for their extreme right-wing views, but they also receive substantial contributions from a number of conventional corporations ranked high in the *Fortune* 500 and have close connections with universities such as Harvard, Stanford, and Washington University.[49] The expertise offered by professors who work with the conservative foundations that have emerged since the early 1970s advocates using state resources to rebuild the power of the private sector in the United States, directing resources from social services toward production. With regard to foreign policy, these professors generally seek American supremacy, in terms of global economic competition and military might. They want to use legislative initiatives, such as tax breaks, free trade, and greatly increased defense spending, to reach these ends.

What has occurred over the past twenty years is the politicization of expertise. Professors, acting not as citizens exercising their civil liberties but as experts offering science-based knowledge, present policy makers with very different alternatives. The difficulties the politicization of expertise presents for the PMC are obvious. When science no longer speaks with a single voice, the ideological dimensions of technical solutions become highly visible. Although this creates problems for professionals, they cannot return to the status quo ante because the disciplines, the state apparatuses, and the idea structures they support are pervaded by factional struggles for hegemony.

When resources for the postsecondary sector contract unevenly, conflict becomes more acute. Max Weber said the following when discussing crises faced by occupational or status groups, such as the professorate: "When the basis of the acquisition and distribution of goods are relatively stable, stratification by status is favored. Every technological repercussion and economic transformation threatens stratification by status and pushes the class situation into the foreground."[50]

The stability of the period prior to Vietnam War contributed to stratification by status, which meant the ideology of expertise developed by faculty was largely unquestioned. Currently, faculty are confronting the major technological and economic transformations described in the Forum reports. The class situation they must deal with is whether or not to align with the institutional class. They cannot return to their previous situation because the ideology of expertise no longer veils the class relations and embodied in the knowledge they deploy outside the university.

The professorate and the PMC face yet another crisis of legitimacy.[51] The ideology being developed by the Forum offers some segments of the PMC an ideology that might alleviate the crisis of legitimacy. The Forum is constructing and entrepreneurial ideology of expertise, one that justifies expertise on the basis of its contribution to commercial success. Although the entrepreneurial ideology being constructed by the Forum obviously serves the institutional class, the project which the ideology proposes—a strong economy and a high standard of living—also speaks to the general welfare. Professionals, who must keep their concern for the general welfare in the public eye in order to legitimate their monopolies of expertise, could conceivably justify their privileged status by embracing an entrepreneurial expertise aimed at recovering standards of prosperity associated with the 1950s and 1960s.

An entrepreneurial ideology of expertise is most likely to be internalized and strongly promoted by faculty at prestigious institutions whose expertise and ambition coincide with the enterprises in which the institutional class is engaged. Their promulgation of an entrepreneurial ideology of expertise will enhance its legitimacy, making it available to faculty throughout postsecondary education who are able to share in activities flowing from business–higher education partnerships. Faculty in the sciences, high technology areas, and business and management schools, from elite research institutions to community colleges, will find it possible to espouse such an ideology because it will legitimate their economic activity outside the university, so long as that activity is within their area of expertise and can be construed as contributing to national economic growth and prosperity.

An entrepreneurial ideology of expertise may temporarily mitigate the crisis of legitimacy facing the PMC by seeing professional expertise as contributing to international economic recovery. However, in the long run is unlikely to serve as a workable set of beliefs. The ideology claims to promote prosperity for society as a whole, but it is actually formed in complementarity with the institutional class and will channel professional activity into promoting that class's projects, the benefits of which will not fall evenly across the citizenry, finally bringing into question the ideology as a whole.

If the ideology is detached from the institutional class yet embraced by faculty, it is likely to foster unbridled competition, pitting faculty against private sector enterprises as well as against each other. As cut-throat competition and the laissez-faire ideology that supported it were checked in the last quarter of the nineteenth century by state regulation, so actively entrepreneurial faculty are likely to be constrained in the last decade of the twentieth century. Indeed, in some states professors have already been asked to declare annually and publicly all their outside income.

Another difficulty with an entrepreneurial ideology is that it will put faculty into an adversarial relationship with the leaders of the universities that are the basis of their support. It is likely to channel the vigor and initiative of the faculty into activity external to their institutions. The scholarship, research and development, and profits produced by their activity will be lost to the university. In such circumstances, institutional managers will probably move to control professional labor more closely, encroaching on discretionary time or exacting a share of the professionals' profits. Many medical practice plans already embody the later and are easily generalizable to other faculty activities.

Despite the appeal of an entrepreneurial ideology of expertise, some professors will probably continue to construct advocacy ideologies of expertise, which value knowledge for what it can contribute to changing existing patterns of privilege. A major difficulty confronting the widespread adoption of such ideologies by faculty is the many and contending visions of a just society. Factional strife has always characterized the academic left, and has been an obstacle to the articulation of hegemonic principles. Another problem confronting proponents of advocacy ideologies of expertise is their dependence on relatively unpredictable coalitions of groups in the wider society for maintaining pressure on the PMC and the institutional class to keep the footholds they have gained in terms of state resources. Without a relatively predictable flow of resources, this fraction of the PMC will not have a base from which to develop an ideology of expertise.

In sum, the PMC faces a crises of legitimacy and is being pressed by organizations such as the Forum to develop an ideology of expertise in complimentarity with the institutional class. If the PMC subscribes to an entrepreneurial ideology of expertise, it will serve to reproduce the culture and class relations of the institutional class. If it subscribes to advocacy ideologies of expertise, it will have to develop notions of legitimate knowledge in complimentarity with subordinate classes.

Both entrepreneurial and advocacy ideologies of expertise assume that the PMC is a derivative class, unable to assert consistently interests of its own. However, the PMC may be able to enhance its autonomy by developing an ideology which will unify and strengthen itself. The hegemonic principles of a new PMC ideology would have to address the uses of knowledge, or expertise, since knowledge is the raison d'être of the PMC. To claim legitimacy, the PMC would have to acknowledge the ideological as well as the technical dimension of expertise. The success of the PMC in creating widespread legitimacy would probably depend on its ability to deal explicitly with the ways in which knowledge intersects with and incorporates power, as well as with class and political interests. The PMC would have to develop a history of concrete mediations in which expertise was deployed to the benefit of various and contending groups or classes. This approach to expertise is not the "old" ideology of expertise revitalized. Rather than claiming that knowledge is objective, value free, above class or partisan struggles, as did the early PMC ideology of expertise, the new ideology would have to recognize that knowledge cannot escape the conditions of the class society in which it is created, and would have to work from there to establish ways in which expertise can contribute to policy formation. Although class unity and ideological clarity are unlikely to bring major social changes, the PMC might contribute to a gradual realignment of class relations by judiciously positioning itself between capital and labor.

The several ideologies of expertise that I have suggested are available to professionals in the last quarter of the twentieth century may not be the only alternatives or even the major ones. The particular ideology is not so important as is our recognition that expertise, which is no more than knowledge deployed in concrete social situations, has ideological dimensions, with political ramifications. Unless we can begin to look at the broad social dynamics that shape the ideology of professionals, we are likely to be able to understand how higher education policy is formed.

Notes

Chapter 1. The Higher Education Policy Literature

1. See, for example, Theodore Caplow and Reece J. McGee, *The Academic Market Place* (New York: Basic Books, 1958), and James R. Mingle et al., *Challenges of Retrenchment* (San Francisco: Jossey-Bass, 1981).

2. For the state as prime funder of higher education, see, for example, Chester E. Finn, Jr., *Scholars, Dollars and Bureaucrats* (Washington, D.C.: Brookings Institute, 1978); and Bruce L.R. Smith and Joseph K. Karlesky, eds., *The State of Academic Science: Background Papers* (New Rochelle, NY: Change Magazine Press, 1978). For the state as an intrusive force, see Robert O. Berdahl, *Statewide Coordination of Higher Education* (Washington, D.C.: American Council on Education, 1975).

3. For some creative work along these lines, see Martin Carnoy, *The State and Political Theory* (Princeton, NJ: Princeton University Press, 1984), and Theda Skocpol, "Bringing the State Back In: Strategy of Analysis in Current Research, " in *Bringing the State Back In*, ed. Peter B. Evans, Dietrich Rueschemeyer, and Theda Skocpol (Cambridge: Cambridge University Press, 1985): 3-37.

4. President's Commission on Industrial Competitiveness, *Global Competition: The New Reality* (Washington, D.C.: USGPO, 1985).

5. Department of Defense, *The Department of Defense Report on the University Role in Defense Research and Development* (Washington, D.C.: The Pentagon, April 1987).

6. Robert Arnove, ed., *Philanthropy and Cultural Imperialism: The Foundations at Home and Abroad* (Boston: G.K. Hall, 1980).

7. Steven Bailey, *Educational Interest Groups in the Nation's Capital* (Washington, D.C.: ACE, 1975); Lauriston King, *The Washington Lobbyists for Higher Education* (Lexington, MA: Lexington Books, 1975).

8. Daniel Greenberg, *The Politics of Pure Science: An Inquiry into the Relationship between Science and Government in the United States* (New York: New American Library, 1967).

9. Paul Dressel, Lewis Mayhew, Cameron Fincher, and Paul Peterson, as quoted in Leif S. Hartmark and Edward R. Hines, "Politics and Policy in Higher Education: Reflections on the Status of the Field," in *Policy Controversies in Higher Education*, ed. Samuel K. Gove and Thomas M. Stauffer (New York: Greenwood, 1986), 4.

10. David D. Henry, *Challenges Past, Challenges Present: An Analysis of American Higher Education since 1930* (San Francisco: Jossey-Bass, 1975).

11. Committee on Financing Higher Education, *Final Report: Nature and Needs of Higher Education* (New York: Columbia University Press, 1952). See also A.D. Henderson, "Contrasting Principles in Higher Education," *School and Society* 77 (March 28, 1953).

12. Henry, *Challenges Past, Challenges Present*; Janet Kerr, "From Truman to Johnson: Ad Hoc Policy Formation in Higher Education," *The Review of Higher Education* 8 (Fall 1984): 15-44. Janet Rogers-Clarke Johnson and Laurence R. Marcus, *Blue Ribbon Commissions and Higher Education: Changing Academe from the Outside* (Washington, D.C.: ASHE-ERIC Higher Education Report No. 2, Association for the Study of Higher Education, 1986).

13. Committee for Economic Development, *The Management and Financing of Colleges* (New York: Committee for Economic Development, October 1973).

14. Carnegie Commission on Higher Education, *Higher Education: Who Benefits: Who Should Pay?* (New York: McGraw-Hill, 1973); Carnegie Commission on Higher Education, *The More Effective Use of Resources: An Imperative for Higher Education* (New York: McGraw-Hill, 1972); Carnegie Council on Policy Studies in Higher Education, *The Federal Role in Postsecondary Education: Unfinished Business, 1975-1980* (San Francisco: Jossey-Bass, 1973).

15. David W. Breneman and Chester E. Finn, Jr., *Public Policy and Private Higher Education* (Washington, D.C.: Brookings Institute, 1978). Bruce K. Maclaury, President, Brookings, "Foreward," viii.

16. Carnegie Council, *A Classification of Institutions of Higher Education* (Washington, D.C.: The Carnegie Foundation for the Advancement of Teaching, 1976, rev. ed.)

17. Lawrence E. Gladieux and Thomas R. Wolanin, *Congress and the Colleges: The National Politics of Higher Education* (Lexington, MA: Lexington Books, 1975).

18. Sloan Commission, *A Program for Renewed Partnership* (Cambridge: Ballinger, 1980).

19. Gender discrimination and affirmative action are a case in point. See Cynthia Fuchs Epstein, *Women's Place: Options and Limits in Professional Careers* (Berkeley: University of California Press, 1970); Alice S. Rossi and Ann Calderwood, eds., *Academic Women on the Move* (New York: Russell Sage Foundation, 1973); Blanche Fitzpatrick, *Women's Inferior Education: An Economic Analysis* (New York: Praeger, 1976); Judith Newcombe and Clifton F. Conrad, "A Theory of Mandated Change," *Journal of Higher Education* 52 (November/December 1981): 555–577; Patricia Hyer, "Affirmative Action for Women Faculty: Case Studies of Three Successful Institutions," *Journal of Higher Education* 56 (May/June 1985): 282–299.

20. Clark Kerr, *The Uses of the University* (Cambridge: Harvard University Press, 1963); Christopher Jencks and David Riesman, *The Academic Revolution* (New York: Doubleday, 1968).

21. For the development of higher education as a field of study, see Paul Dressel and Lewis Mayhew, *Higher Education as a Field of Study* (San Francisco: Jossey-Bass, 1974). The major associations for higher education are the American Association of Higher Education, the American Educational Research Association, Division J, and the Association for the Study of Higher Education. The central journals are *The Journal of Higher Education, Higher Education, The Review of Higher Education,* and *Research in Higher Education.*

22. Gladieux and Wolanin, *Congress and the Colleges*: King, *The Washington Lobbyists for Higher Education*. In the preface and introductory materials, the authors of these volumes did not acknowledge organizational help or sponsorship of any kind. However, the lack of acknowledgement cannot be seen as a definitive statement of lack of sponsorship.

23. This list was compiled by examining the prefatory and introductory materials in the major works of these authors in the Graduate Library at the University of Arizona, and of course is not exhaustive. It does, however, give a suggestion of policy sponsors.

24. Gladieux and Wolanin, *Congress and the Colleges.*

25. Barbara Ann Scott, *Crisis Management in American Higher Education* (New York: Praeger, 1983).

26. On behavioralism, see David M. Ricci, *The Tragedy of Political Science: Politics, Scholarship and Democracy* (New Haven, CT: Yale University Press, 1984), especially Chapter 5, "The Behavioral Persuasion."

27. David Dickson, *The New Politics of Science* (New York: Pantheon, 1984).

28. A. Hunter Dupree, *Science in the Federal Government: A History of Policies and Activities* (Cambridge: Belknap Press of Harvard University Press, 1957).

29. Dickson, *The New Politics of Science*.

30. National Science Foundation, *Corporate Science: A National Study of University and Industry Researchers* (Washington, D.C.: National Science Foundation, 1984); Seymour Melman, *The Permanent War Economy: American Capitalism in Decline* (New York: Simon and Schuster, 1974).

31. National Science Board, *University-Industry Research Relationships* (Washington, D.C.: National Science Foundation, 1982); Thomas W. Langfitt et al., eds., *Partners in the Research Enterprise: University-Corporate Relationships* (Philadelphia: University of Pennsylvania Press, 1983); Bernard D. Reams, Jr., *University-Industry Research Partnerships: The Major Legal Issues in Research and Development Agreements* (Westport, CT: Quoram, 1986).

32. Business–Higher Education Forum, *Corporate and Campus Cooperation: An Action Agenda* (Washington, D.C.: Business–Higher Education Forum, May 1984), p. 11.

33. Lynn G. Johnson, *The High Technology Connection: Academic/Industrial Cooperation for Economic Growth* (Washington, D.C.: Association for the Study of Higher Education, 1984; ASHE-ERIC Higher Education Research Report No. 6); Robert F. Johnston and Christopher G. Edwards, *Entrepreneurial Science: New Links Between Corporations, Universities and Government* (Westport, CT: Greenwood, 1987); U.S. Congressional Office of Technology Assessment, *Technology, Innovation, and Regional Economic Development* (Washington, D.C.: U.S. Congressional Office of Technology Assessment, 1984); U.S. Department of Commerce, *High Technology Industries: Profiles and Outlooks—Biotechnology* (Washington, D.C.: Department of Commerce, July 1984).

34. Most claims about the contribution of science to innovation are made on the basis of a few studies. However, science does not necessarily mean university research, and the payoffs in terms of product development for universities have not been great in recent years. See Martin Kenney, *Biotechnology: The University-Industrial Complex* (New Haven, CT: Yale University Press, 1986), and Rikard Stankiewicz, *Academics and Entrepreneurs: Developing University-Industry Relations* (London: Frances Pinter, 1986).

35. James R. O'Connor, *The Meaning of Crisis: A Theoretical Introduction* (New York: Basil Blackwell, 1987); Barry Bluestone and Bennett Harrison, *Plant Closings, Community Abandonment and the Dismantling of Basic Industry* (New York: Basic Books, 1982).

36. This point has been debated by the American Association of State Colleges and Universities, which is trying to create a role for nonelite four-year institutions in economic development. See American Association of State

Colleges and Universities, *The Higher Education–Economic Development Connection: Emerging Roles for Public Colleges and Universities in a Changing Economy* (Washington, D.C.: AASCU, 1986).

37. Kenney, *Biotechnology: The University-Industrial Complex.*

38. Derek Curtis Bok, *Beyond the Ivory Tower: Social Responsibilities of the Modern University* (Cambridge: Harvard University Press, 1982); Donald Kennedy, Evidence in U.S. Congress, House Committee on Science and Technology, Subcommittee on Investigations and Oversight, Subcommittee on Science, Research, and Technology, *Commercialization of Academic Biomedical Research* (Hearings, 97th Congress, 1st Session, June 8–9, 1981), (as quoted in Dickson, *The New Politics of Science, 8).*

39. See, for example, Association of American Universities, *University Policies on Conflict of Interest and Delay of Publication* (Washington, D.C.: AAU, February 1985).

40. Thorstein Veblen, *The Higher Learning in America: A Memorandum on the Conduct of Universities by Business Men* (New York: Viking Press, 1918); Washington Gladden, "Tainted Money," *New Outlook* 52 (1895):886–887. See also Robert H. Bremner, *American Philanthropy* (Chicago: University of Chicago Press, 1960).

41. Upton Sinclair, *The Goosestep, A Study of American Education* (Pasadena, CA: The Author, 1923); Ernest Victor Hollis, *Philanthropic Foundations and Higher Education* (New York: Columbia University Press, 1938); Hubert Park Beck, *Men Who Control Our Universities: The Economic and Social Composition of Governing Boards of Thirty Leading American Universities* (New York: King's Crown Press, 1947); James Ridgeway, *The Closed Corporation: American Universities in Crisis* (New York: Random House, 1968); Irving Louis Horowitz, ed., *The Rise and Fall of Project Camelot: Studies in the Relationship Between Social Science and Practical Politics* (Cambridge: MIT Press, 1967); Samuel Bowles and Herbert Gintis, *Schooling in Capitalist America* (New York: Basic Books, 1976); David N. Smith, *Who Rules the Universities* (New York: Monthly Review Press, 1974).

42. See, for example, E.C. Wallenfeldt, *American Higher Education: Servant of the People or Protector of Special Interests?* (Westport, CT: Greenwood, 1983).

43. Although the critical tradition has not flourished in the postsecondary literature, it is alive and well at the schools level, where it incorporates many of the debates central to neo-Marxian conceptual developments with regard to class, state, and ideology. See, for example, Michael W. Apple, *Ideology and Curriculum* (London: Routledge and Kegan Paul, 1979); Pierre Bourdieu and Jean-Claude Passeron, *Reproduction in Education, Society and Culture* (London

and Beverly Hills: Sage, 1977); Pierre Bourdieu and Jean-Claude Passeron, *The Inheritors: French Students and Their Relation to Culture* (Chicago: University of Chicago Press, 1979); Martin Carnoy and Henry M. Levin, *Schooling and Work in the Democratic State* (Stanford, CA: Stanford University Press, 1985); Lois Weis, *Between Two Worlds: Black Students in an Urban Community College* (Boston: Routledge and Kegan Paul, 1985); Paul Willis, *Learning to Labor: How Working Class Kids Get Working Class Jobs* (New York: Columbia Press, 1981).

44. Warren O. Hagstrom, *The Scientific Community* (New York: Basic Books, 1965); Diane Crane, "Scientists at Minor and Major Universities: A Study of Productivity and Recognition," *American Sociological Review* 30 (October 1965):699–714; Warren O. Hagstrom, "Inputs, Outputs and the Prestige of University Science Departments," *Sociology of Education* 44 (1971):375–397; Stephen Cole and Jonathan R. Cole, "Scientific Output and Recognition: A Study in the Operation of the Reward System in Science," *American Sociological Review* 32 (June 1967):377–390; Warren O. Hagstrom, "Competition in Science," *American Sociological Review* 39 (February 1974):1–18; Paul I. Allison and John A. Stewart, "Productivity Difference Among Scientists: Evidence for Accumulation Advantage," *American Sociological Review* 39 (August 1974):596–606.

45. Dael Wolfle, *Science and Public Policy* (Lincoln: University of Nebraska Press, 1977); Dael Wolfle, Carnegie Commission Report, *The Home of Science: The Role of the University* (New York: McGraw-Hill, 1972): Smith and Karlesky, *The State of Academic Science.*

46. Thomas J. Kuehn and Alan L. Porter, eds., *Science, Technology and National Policy* (Ithaca, NY: Cornell University Press, 1981); Malcolm L. Goggin, *Governing Science and Technology in a Democracy* (Knoxville: University of Tennessee Press, 1986).

47. Dickson, *The New Politics of Science*; Greenberg, *The Politics of Pure Science*; Kenney, *Biotechnology: The University-Industrial Complex*; Sheldon Krimsky, *Genetic Alchemy* (Cambridge: MIT Press, 1982); Dorothy Nelkin, *Science as Intellectual Property* (New York: Macmillan, 1984); David Noble, *America by Design: Science, Technology and The Rise of Corporate Capitalism* (New York: Knopf, 1976).

48. John T. Wilson, *Academic Science Higher Education and the Federal Government 1950-1983* (Chicago: University of Chicago Press, 1983).

49. As quoted in Business–Higher Education Forum, *Corporate and Campus Cooperation*, 5.

50. National Academy of Engineering, *The Technological Dimensions of International Competitiveness* (Washington, D.C.: National Academy of Engineering, 1988). U.S. Congressional Office of Technology Assessment, *Technology In-*

novation and Regional Economic Development (Washington, D.C.: U.S. Congressional Office of Technology Assessment, 1984).

51. Business-Higher Education Forum, Handout, December 17, 1987, 5.

52. Ibid., 3-4.

53. An institution was judged selective if it was defined as such in Alexander W. Astin, *Predicting Academic Performance in College: Selectivity by Data for 2300 American Colleges* (New York: Free Press, 1971); in Category 3.1 in Carnegie Commission on Higher Education, *A Classification of Institutions of Higher Education*; or in Thomas R. Dye's list of 25 private universities with the largest endowments, which appears in his *Who's Running America? The Conservative Years* (Englewood Cliffs, NJ: Prentice-Hall, 1986, 4th ed.).

54. Business-Higher Education Forum, Handout, December 17, 1987, 5.

Chapter 2. Policy Issues Past and Present

1. John Kenneth Galbraith *The New Industrial State* (Boston: Houghton Mifflin, 1985); Barry Bluestone and Bennett Harrison, *The Deindustrialization of America* (New York: Basic Books, 1982).

2. Dorothy Nelkin, *Science as Intellectual Property: Who Controls Scientific Research* (New York: Macmillan, 1984); Sheila Slaughter, "Academic Freedom and the State: Reflections on the Uses of Knowledge," *Journal of Higher Education* 59 (May/June 1988): 241-267.

3. For neo-Marxian interpretations of the falling rate of profit, see James O'Connor, *The Fiscal Crisis of the State* (New York: St. Martin's Press, 1973); and Joaquim Hirsch, "The State Apparatus and Social Reproduction: Elements of a Theory of the Bourgeois State," in *State Capital: A Marxist Debate*, ed. J. Holloway and Sol Picciotto (London: Edward Arnold, 1985). For more mainstream interpretations, see Galbraith, *The New Industrial State*, and Lester C. Thurow, *The Zero-Sum Solution: Building a World-Class American Economy* (New York: Simon and Schuster, 1985).

4. Howard Bowen, *Investment in Learning: The Individual and Social Value of American Higher Education* (San Francisco: Jossey-Bass, 1977).

5. For changes in student enrollment in degree fields from arts and sciences to technical and professional fields, see U.S. Department of Education, *Digest of Educational Statistics* (Washington, D.C.: Center for Educational Statistics, 1987), and U.S. Department of Education, *The Condition of Education* (Washington, D.C.: Center for Educational Statistics, 1987).

6. For business leaders, see Business-Higher Education Forum, *America's Competitive Challenge: The Need for a National Response* (Washington, D.C.: Business-Higher Education Forum, April 1983); for university presidents, see Derek Curtis Bok, *Beyond the Ivory Tower: Social Responsibilities of the Modern University* (Cambridge: Harvard University Press, 1982); Paul E. Gray's comments in "New President is Chosen at MIT: He Warns of U.S. Technology Lag," *New York Times* (October 6, 1980); Sheldon Hackney, "Prologue," in Thomas W. Langfitt et al., *Partners in the Research Enterprise: University-Corporate Relations in Science and Technology* (Philadelphia: University of Pennsylvania Press, 1983); for political leaders, see Gov. Mario M. Cuomo in "State of the State" address as cited in *The Chronicle of Higher Education* (February 3, 1988, 34:21, p. A28; and Albert Gore, Jr., "Recombined Institutions: The Changing University-Corporate Relationship," in Hackney, *Partners in the Research Enterprise:* 121-127.

7. See, for example, Robert F. Johnston and Christopher G. Edwards, *Entrepreneurial Science: New Links between Corporations, Universities and Government* (Westport, CT: Greenwood, 1987).

8. Burton J. Bledstein, *The Culture of Professionalism* (New York: W. W. Norton, 1977); Magali Sarfatti Larson, *The Rise of Professionalism: A Sociological Analysis* (Berkeley: University of California Press, 1977).

9. Christopher Jencks and David Riesman, *The Academic Revolution* (Garden City, NY: Doubleday, 1968).

10. Ibid.

11. Harry T. Edwards, *Higher Education and the Unholy Crusade Against Government Regulation* (Cambridge: Harvard University, Institute for Educational Management, 1980).

12. O'Connor, The *Fiscal Crisis of the State.*

13. National Science Foundation, *Women and Minorities in Science and Engineering* (Washington, D.C.: NSF, January 1984).

14. Ibid. See also Helen S. Astin and M.B. Snyder, "Affirmative Action 1972-82: A Decade of Response," *Change* 14 (July 1982): 26-31.

15. Elchanan Cohn and Larry L. Leslie, "The Development and Finance of Higher Education in Perspective," in *Subsidies to Higher Education: The Issues*, ed. Howard P. Tuchman and Edward Whalen (New York: Praeger, 1980): 11-32.

16. Judith K. Lawrence and Kennth C. Green, *A Question of Quality: The Higher Education Ratings Game* (Washington, D.C.: American Association for Higher Education, ERIC/Higher Education Research Report No. 5, 1985).

17. National Science Foundation, *Academic Science 1972–1981: Surveys of Science Resources Series, Final Report NSF 81–326* (Washington, D.C.: National Science Foundation, December 1981):8; see also Dael Wolfle, "Forces Affecting the Research Role of Universities," in *The State of Academic Science: Background Papers*, ed. Bruce L.R. Smith and Joseph J. Karlesky (New York: Change Magazine Press, 1978): 17–59.

18. Allan M. Cartter, *An Assessment of Quality in Graduate Education* (Washington, D.C.: American Council on Education, 1966); Kenneth D. Roose and Charles J. Anderson, *Rating of Graduate Programs* (Washington, D.C.: American Council on Education, 1970); Lyle V. Jones, Gardner Lindzey, and Porter E. Coggershall, eds., *An Assessment of Research-Doctorate Programs in the United States* [5 volumes] *Biological Sciences; Engineering; Humanities; Mathematical and Physical Sciences; Social and Behavioral Sciences* (Washington, D.C.: National Academy Press, 1982).

19. Robert Goodman, *The Last Entrepreneurs: America's Regional Wars for Jobs and Dollars* (New York: Simon and Schuster, 1979).

20. Sheila Slaughter and Edward T. Silva, "Toward a Polictical Economy of Retrenchment: The American Public Research Universities," *The Review of Higher Education* 8 (Summer 1985): 295–318.

21. Bluestone and Harrison, *The Deindustrialization of America*.

22. Goodman, *The Last Entrepreneurs*.

23. Bruce L.R. Smith and Joseph J. Karlesky, *The State of Academic Science: The Universities in the Nation's Research Effort* (New York: Change Magazine Press, 1977): 43.

24. Lawrence and Green, *A Question of Quality*, 17. This table constructs rankings for universities by aggregating the departmental reputational scores, which are the basis of the American Council on Education's rankings and therefore should be treated with caution. This table does not include the ACE 1982 ratings, reported in Lindzey and Coggershall, cited in note 18.

25. Larson, *The Rise of Professionalism*; Randall Collins, *The Credential Society* (New York: Academic Press, 1979).

26. Collins, *The Credential Society* Tables 4.1 and 4.2, quote at p. 85.

27. Ibid.

28. Carnegie Council on Policy Studies in Higher Education, *Three Thousand Futures: The Next Twenty Years for Higher Education* (San Francisco: Jossey-Bass, 1980).

29. Ibid.

30. Robert O. Berdahl, *State Wide Coordination of Higher Education* (Washington, D.C.: American Council on Education, 1975).

31. Stephen K. Bailey, *Educational Interest Groups in the Nation's Capital* (Washington, D.C.: American Council on Education, 1975).

32. Paul Dressel and Lewis Mayhew, *Higher Education as a Field of Study* (San Francisco: Jossey-Bass, 1974).

33. Amitai Etzioni, ed., *The Semi-Professions and Their Organization: Teachers, Nurses and Social Workers* (New York: Free Press, 1969).

34. O'Connor, *The Fiscal Crisis of the State.*

35. The volume edited by James R. Mingle et al., *Challenges of Retrenchment* (San Francisco: Jossey-Bass, 1981), is an example of how to implement retrenchment in the area of postsecondary education.

36. Academic Freedom and Tenure, "City University of New York: Mass Dismissals Under Financial Exigency," *American Association of University Professors Bulletin* 63 (April 1977): 60-81.

37. Academic Freedom and Tenure, "The State University of New York," *American Association of University Professors Bulletin* 63 (August 1977): 237-260.

38. United University Professors, Untitled factsheet, Albany, N.Y., 1983. See also Barbara Ann Scott, *Crisis Management in American Higher Education* (New York: Praeger, 1983).

39. Ibid.

40. According to the National Science Foundation, health and national defense, with the exception of a brief period in the mid-1970s, have supplied the largest amounts of money for academic research since regular records were kept. National Science Foundation, *Federal Support to Universities, Colleges and Selected Nonprofit Institutions Fiscal Year 1981* (Washington, D.C.: USGPO, 1981), Table B-2. This trend will be discussed in detail later in this chapter. See also Barbara S. Clowse, *Brainpower for the Cold War: The Sputnik Crisis and the Naitonal Defense Education Act of 1958* (Westport, CT: Greenwood, 1981).

41. James Baxter III, *Scientists Against Time* (Boston: Little Brown, 1946).

42. Herbert F. York and G. Allen Greb, "Military Research and Development: A Postwar History," in *Science, Technology and National Policy,* ed. T. Kuehn and A.L. Porter (Ithaca, NY: Cornell University Press, 1981); David Dickson, *The New Politics of Science* (New York: Pantheon Books, 1984), especially Chapter 3, "Science and the Military: Knowledge as Power."

43. Theodore R. Vallance, "Classified Research and Related Issues in Science Communication," *American Association of University Professors Bulletin* 55 (Autumn 1969): 360–365.

44. Robert A. Rosenbaum et al., "Federal Restrictions on Research: Academic Freedom and National Security," *Academe: Bulletin of the Association of American University Professors* 68 (September–October 1982).

45. Dickson, *The New Politics of Science*.

46. York and Greb, "Military Research and Development."

47. National Science Foundation, *Federal Support to Universities, Colleges and Selected Nonprofit Institutions Fiscal year 1982* (Washington, D.C.: USGPO, 1982), Table B-2.

48. Ibid.

49. Robert M. Rosenzweig with Barbara Turlington, *The Research Universities and Their Patrons* (Berkeley: University of California Press, 1982). Rosenzweig was vice-provost at Stanford, and now heads the Association of American Universities and sits on the Business–Higher Education Forum. See also Bok, *Beyond the Ivory Tower*.

50. Bok, *Beyond the Ivory Tower*; Langfitt et al., *Partners in the Research Enterprise*.

51. Business–Higher Education Forum, *America's Competitive Challenge*, 9.

52. National Science Board, *Basic Research in the Mission Agencies: Agency Perspectives on the Conduct and Support of Basic Research* (Washington, D.C.: National Science Foundation, 1978): 57–58.

53. Academic Freedom and Tenure, "The Enlargement of the Classified Information Systems," *Academe* 69 (January/February 1983): 9a–14a.

54. Christina Ramirez, "The Balance of Interests Between National Security Controls and First Amendment Interests in Academic Freedom" (Houston, TX: Institute for Higher Education Law and Governance, 1985).

55. Randall D. Knight, "Science, Space, and Scholarship: University Research and the Strategic Defense Initiative," *Educational Policy* 1, 4 (1987): 499–512.

56. National Science Foundation, *University-Industry Research Relationships: Myths, Realities and Potentialities* (Washington, D.C.: National Science Foundation, 1982).

57. Descriptions of university-industry agreements abound; among the

better general ones are Lynn G. Johnson, *The High Technology Connection: Academic/Industrial Cooperation for Economic Growth* (Washington, D.C.: Association for the Study of Higher Education, 1984, ASHE-ERIC Higher Education Research Report No. 6); Langfitt, et al., *Partners in the Research Enterprise*; National Science Board, *University Industry Research Relationships* (Washington, D.C.: National Science Foundation, 1982).

58. Dickson, *The New Politics of Science*.

59. Office of Technology Assessment, *Technology, Innovation and Regional Economic Development* (Washington, D.C.: U.S. Congress, Office of Technology Assessment, July, 1984).

60. Bluestone and Harrison, *The Deindustrialization of America*.

61. Gilbert S. Omenn, "University-Corporate Relations in Science and Technology: An Analysis of Specific Models," in Langfitt et al., *Partners in the Research Enterprise*.

62. Nelkin, *Science as Intellectual Property*.

63. Dickson, *The New Politics of Science*, Figure 1.

64. National Science Foundation, *Federal Funds for Research and Development, Fiscal Years 1978, 1979 and 1980*, Vol. 28, *Surveys of Science Resources Series* NSF 80-315 (Washington, D.C.: USGPO, May 1980).

65. National Science Foundation, *Federal Funds for Research and Development, Fiscal Years 1980, 1981, and 1982*, Vol. 30, *Surveys of Science Resources Series* NSF 82-321 (Washington, D.C.: USGPO, April 1982).

66. National Science Foundation, *Science Resources Studies Highlights* NSF 84-333 (Washington, D.C.: USGPO, November 30, 1984).

67. Margaret Rossiter, *Women Scientists in America* (Baltimore, MD: Johns Hopkins University Press, 1982).

68. This overview of the Forum's legislative agenda, which is discussed in detail in Chapters 6 and 7, is based on the following Business–Higher Education Forum reports: *America's Competitive Challenge: The Need for a National Response* (Washington, D.C.: B-HEF, April 1983); *Corporate and Campus Cooperation: An Action Agenda* (Washington, D.C.: B-HEF, May 1984); *The New Manufacturing: America's Race to Automate* (Washington, D.C.: B-HEF, June 1984); *Space: America's New Competitive Frontier* (Washington, D.C.: B-HEF, April 1986); *Export Controls: The Need to Balance National Objectives* (Washington, D.C.: B-HEF, January 1986); *An Action Agenda for American Competitiveness* (Washington, D.C.: B-HEF, September 1986). Dickson, *The New Politics of Science*, also sees these points as crucial to the Forum's legislative agenda.

69. Business-Higher Education Forum, *America's Competitive Challenge:* 2.

70. For the effect of patterns of pooling on competition, see Gabriel Kolko, *Railroads and Regulation, 1877-1916* (New York: W.W. Norton, 1970), and William Appleman Williams, *Contours of American History* (Chicago: Quadrangle, 1966).

71. Willard W. Lawrence, "Choosing Our Pleasures and Our Poisons: Risk Assessment for the 1980s," in *Science, Technology and the Issues of the Eighties: Policy Outlook*, eds. Albert H. Teich and Ray Thornton. (Boulder, CO: Westview Press, 1982): 99-120; Rosalie Bertell, *No Immediate Danger? Prognosis for a Radioactive Earth* (Toronto: Women's Educational Press, 1982).

72. On patent litigation and its economic consequences, see David F. Noble, *Forces of Production: A Social History of Industrial Automation* (New York: Knopf, 1984).

73. Business-Higher Education Forum, *America's Competitive Challenge:* 1.

74. Ibid.: 2.

75. See, for example, Lawrence Gladieux and Thomas Wolanin, *Congress and the Colleges: The National Politics of Higher Education* (Lexington, MA: Lexington Books, 1976).

76. See, for example, Thomas J. Kuehn and Alan L. Porter, eds., *Science, Technology and National Policy* (Ithaca, NY: Cornell University Press, 1981).

77. Samuel Bowles and Herbert Gintis, *Schooling in Capitalist America* (New York: Basic Books, 1976).

78. Martin Carnoy and Henry M. Levin, *Schooling and Work in the Democratic State* (Palo Alto, CA: Stanford University Press, 1985).

79. In the main, science policy has not been explicitly analyzed using neo-Marxist theory focused on the present. As indicated in Chapter 1, note 41, neo-Marxist analysis has concentrated on the schools in the educational sector.

80. Gerhard Lehmbruch and Philippe Schmitter, eds., *Patterns of Corporatist State Policy Making* (London: Sage, 1982); Leo Panitch, "Recent Theorizations of Corporatism," *British Journal of Sociology* (June 1974): 159-182.

81. There is little corporatist theory in the United States directed at postsecondary education. The Forum perhaps comes as close to developing such a strand of analysis as any work thus far.

Chapter 3. Social Location of Corporate and University Leaders

1. See, for example, Edward Banfield, *Political Infleunce* (New York: Free Press, 1961); Robert Dahl, *Who Governs? Democracy and Power in an American City* (New Haven, CT: Yale Unviersity Press, 1961); and Charles Lindblom, *The Intelligence of Democracy* (New York: Free Press, 1965).

2. Early pluralist theorists sometimes developed into neopluralist theorists when meeting the critiques, especially from the left, of their work. See, for example, Robert Dahl, *Dilemmas of Pluralist Democracy: Autonomy versus Control* (New Haven, CT: Yale University Press, 1982); Charles Lindblom, "Still Muddling, Not Yet Through," *Public Administration Review* 39 (November/December 1979):517–526. See also Nelson Polsby, *Community Power and Political Theory: Problems of Evidence and Inference* (New Haven, CT: Yale University Press, 1980, 2nd ed.); Patrick Dunleavy and Brendan O'Leary, *Theories of the State: The Politics of Liberal Democracy* (New York: Meredith Press, 1987). For neopluralist theory, see also John Kenneth Galbraith, *The New Industrial State* (Boston: Houghton Mifflin, 1985).

3. For a summary of classical Marxian theory and the state, see Martin Carnoy, *The State and Political Theory* (Princeton, NJ: Princeton University Press, 1984), especially Chapter 2, "Marx, Engles, Lenin and the State."

4. G. William Domhoff, *The Higher Circles* (New York: Random House, 1970); G. Wililam Domhoff, *The Powers That Be: Processes of Ruling Class Domination in America* (New York: Random House, 1978); Ralph Milliband, *The State and Capitalist Society* (London: Weidenfeld and Nicholson, 1969). There are also a number of American theorists who see corporate-based policy actors with business interests broad enough to be called class interests, but are unspecific about their relation to an upper or ruling class. See, for example, C. Wright Mills, *The Power Elite* (New York: Oxford, 1956); James O'Connor, *The Fiscal Crisis of the State* (New York: St. Martin's Press, 1973): Michael Useem, *The Inner Circle: Large Corporations and the Rise of Business Political Activity in the U.S. and U.K.* (New York: Oxford University Press, 1984).

5. Val Burris, "Structuralism and Marxism," *The Insurgent Sociologist* 9 (Summer 1979): 4–17; Martin Carnoy, *The State and Political Theory*, especially Chapter 4, "Structuralism and the State: Althusser and Poulantzas"; Nicos Poulantzas, *Political Power and Social Classes* (London: New Left Books, 1974).

6. Philippe Schmitter, "Still the Century of Corporatism?" *Review of Politics* 36 (January 1974): 85–131; Leo Panitch, "Recent Theorizations of Corporatism," *British Journal of Sociology* (June 1974): 159–182; Gerhard Lehmbruch and Philippe Schmitter, eds., *Patterns of Corporatist State Policy Making* (London: Sage, 1982).

7. G. William Domhoff, ed., *Power Structure Research* (Beverly Hills, CA: Sage, 1980).

8. Many of these men and women are still members, but there has been some change over time.

9. *Who's Who in America* (Chicago: Marquis, 1985); *Who's Who in Finance and Industry* (Chicago, Marquis, 1983); *Who's Who in Education* (Cambridge: International Who's Who in Education, 1981); *Who's Who of American Women* (Chicago: Marquis, 1978).

10. Rosalind Rosenberg, *Beyond Separate Spheres: Intellectual Roots of Modern Feminism* (New Haven: Yale, 1982) speaks to women's confinement to separate spheres in higher education and their early attempts to move beyond them.

11. At this point, I am using a mechanical definition of class, and see upper-class membership as signified (1) by inclusion of the person or his or her parents in the *Social Register*, (2) by attendance by the person or his or her parent at a distinguished private preparatory school, (3) by membership on the part of the individual or his or her parents in one of the very exclusive "gentlemen's" social clubs, (4) if the person's father was the CEO of a large corporatoin, or head of a very large law firm, or (5) if the person marries someone with any of the above characteristics. This definition is taken from G. William Domhoff, *Who Rules America?* (Englewood Cliffs, NJ: Prentice-Hall, 1967). Domhoff's upper class is defined primarily through social characteristics and activity in the corporate sector. I am using Thomas R. Dye, *Who's Running America: The Conservative Years* (Englewood Cliffs, NJ: Prentice-Hall, 1986, 4th ed.), to define elites. He sees elites as those who hold the top positions of power in a variety of sectors: corporate, public interest (mass media, education, foundations, law, civic and cultural organizations), and government (legislative, executive, judicial branches as well as the military). By and large he does not deal with the issue of class as a heritable, social phenomenon and looks instead to the institutional basis of power. For my purposes, if the parent of one of the leaders occupied one of these positions, it would indicate that their family of origin was a part of a national elite. For the person's family of origin to be upper class, it would have to exhibit the characteristics Domhoff describes, although overlaps between Dye's and Domhoff's criteria occur at the point of corporate leadership.

12. Carnegie Council, *A Classification of Institutions of Higher Education* (Washington, D.C.: The Carnegie Foundation for the Advancement of Teaching, 1976, rev. ed.)

13. Carnegie Council, *A Classification of Institutions of Higher Education*. Universities were classified as Research 1.2 if they were on the list of the 100 leading institutions in terms of federal financial support, awarded at

least 50 Ph.D.'s, or were among the leading 60 institutions in terms of total number of Ph.D.'s awarded.

14. Thomas R. Dye, *Who's Running America? The Carter Years* (Englewood Cliffs, NJ: Prentice-Hall 1979, 2nd ed.). For a list of these institutions, see Table 5.3, p. 134. Many overlap with Ivy League institutions.

15. Council of Graduate Schools, *The Master's Degree* (Washington, D.C.: CGS, 1975, rev. ed.).

16. I tried to distinguish earned from honorary degrees, although this was sometimes difficult, especially in the case of corporate leaders. If a Ph.D. was not taken fairly quickly after completion of the B.A. or M.A., allowing time for military service when this seemed indicated, the degree was classified as honorary.

17. Herbert Park Beck, *Men Who Control Our Universities* (New York: King's Crown Press, 1947).

18. John S. Brubacker and Willis Rudy, *Higher Education in Transition: A History of American Colleges and Universities 1636-1976* (New York: Harper and Row, 1976, 3rd ed., revised and enlarged).

19. David Featherman and Robert Hauser, *Opportunity and Change* (New York: Academic Press, 1978).

20. Dye, *Who's Running America? The Conservative Years* (Englewood Cliffs, NJ: Prentice Hall, 1986 4th ed.): 189.

21. William H. Sewell and Robert M. Hauser, *Education, Occupation, and Earnings: Achievement in the Early Career* (New York: Academic Press, 1975).

22. Randall Collins, *The Credential Society: An Historical Sociology of Education and Stratification* (New York: Academic Press, 1979) especially pp. 43-48, "Cultural Credentials and Mobility Barriers."

23. Useem, *The Inner Circle*, p. 72.

24. Mark Green and Andrew Buchsbaum, "The Corporate Lobbies: The Two Styles of the Business Roundtable and Chamber of Commerce," in *The Big Business Reader*, ed. Mark Green et al., (New York: The Pilgrim Press, 1983), 207.

25. G. William Domhoff, *The Powers that Be: Processes of Ruling-Class Domination in America* (New York: Vintage, 1979), 69-70.

26. G. William Domhoff, *The Powers that Be*, 64-67; Laurence H. Shoup and William Minter, *Imperial Brain Trust* (New York: Monthly Review Press, 1977).

27. Reporting of regional association memberships should probably be treated with even more caution than reporting of national memberships. The *Who's Who* series is national in scope; thus, regional memberships are likely to be underreported.

28. The calculation of membership in higher education associations illustrates again the reportage problems with the *Who's Who* data base. Only four (10 percent) of the presidents reported Association of American University membership, but it is clear that approximately 60 percent are members.

29. Several studies have developed lists of social clubs that indicate elite status. Among these are Dye, *Who's Running America? The Conservative Years*, which uses a list provided courtesy of Domhoff, *Who Rules America?* and Useem, *The Inner Circle*. I used Useem here because his list is more recent and shorter, using 15 clubs rather than 36.

30. Michael D. Cohen and James G. March, *Leadership and Ambiguity: The American College President* (Boston: Harvard Business School Press, 1986, 2nd ed.); Thomas R. Dye, *Who's Running America? The Carter Years* (Englewood Cliffs, NJ: Prentice-Hall, 1979, 2nd edition); Thomas R. Dye, *Who's Running America? The Conservative Years* (Englewood Cliffs, NJ: Prentice-Hall 1986, 4th edition); Michael Useem, *The Inner Circle* (New York: Oxford, 1984).

31. The four studies are Frederick de W. Bolman, *How College Presidents are Chosen* (Washington, D.C.: American Council on Education, 1967); Nicholas J. Demerath, Richard W. Stephens, and R. Robb Taylor, *Power, Presidents, and Professors* (New York: Basic Books, 1967); Michael R. Ferrari, *Profiles of American College Presidents* (East Lansing, MI: Michigan State University Business School, 1970); Mark H. Ingraham, *The Mirror of Brass: The Compensation and Working Conditions of College and University Adminstrators* (Madison, WI: University of Wisconsin Press, 1965).

32. Cohen and March, *Leadership and Ambiguity*, Chapter 2, "Prepresidential Careers," 7–28.

33. Clark Kerr and Marian L. Gade's *The Many Lives of Academic Presidents: Time, Place and Character* (Washington, D.C.: Association of Governing Boards, 1986) is more recent, but does not concentrate on research universities nor offer much more detail about the social location from which presidents come.

34. Cohen and March, *Leadership and Ambiguity*, 12.

35. Ibid., 15.

36. Ibid., 8.

37. Ibid., 7–8.

38. In the Forum sample, I counted M.A.'s (15) and Ph.D.'s (7) separately, for a total of 22. This may involve double counting and may exaggerate the number of advanced degrees. It is not clear how Dye computed his advanced degrees.

39. I used the B.A. rather than advanced degrees to determine institutional control.

40. The indicators Dye used for upper-class social origin were roughly the same as those discussed in note 11.

41. Michael Patrick Allen, "Economic Interest Groups and the Corporate Elite Structure," *Social Science Quarterly* 58 (March 1978): 597–615.

42. Useem, *The Inner Circle*. Useem drew his sample from the top and botton of the *Fortune 500 Industrial* and *Fortune Service 500*. As detailed in Chapter 5, Forum presidents are more likely to be at the top of the 500s.

43. On the one hand, there may be a reportage problem for federal government service on the part of CEOs; they may not consider it significant enough to list in what is basically a short vita. On the other hand, when CEOs were checked against Ronald Brownstein and Nina Easton, *Reagan's Ruling Class: Portraits of the President's Top One Hundred Officials* (New York: Pantheon, 1983), which included many of the advisory boards, none of the CEOs appeared.

44. Useem, *The Inner Circle*, 61.

45. Ibid., 41, 61.

46. Dye, *Who's Running America?: The Conservative Years*, 4th Ed.

47. Of course, universities are closely connected to the state, and are in many instances state institutions. However, administrators do not think of themselves of state functionaries, a point I will return to in the final chapter.

Chapter 4. University Presidents and Public Policy

1. See for example, Charles Lindblom, *The Intelligence of Democracy* (New York: Free Press, 1965) for pluralist theory of policy making, and H. Heclo and A. Wildavsky, *The Private Government of Public Money* (London: Macmillan, 1974. 2nd ed.) and Robert Dahl, *Dilemmas of Pluralist Democracy: Autonomy vs. Control* (New Haven: Yale University Press, 1982) for neopluralist positions.

2. For the critical role of the dominant class, despite fluid class structures, see Eric Olin Wright *Class, Crisis and the State* (London: New Left Books, 1978); for the hotly debated relation of the middle class to the dominant class see Pat Walker, ed. *Between Labor and Capital* (Boston: South End Press, 1979).

3. For corporatist positions on policy making see Leo Panitch, "Recent Theorizations of Corporatism: Reflections on a Growth Industry," *British Journal of Sociology 31* (June 1980): 159-187; and Gerhard Lehembruch and Philippe C. Schmitter, eds., *Patterns of Corporatist Policy Making* (London: Sage, 1982).

4. Association of American Universities, Untitled, January 1986, p. 1.

5. For a review of the standard indicators of excellence, see Judith K. Lawrence and Kenneth C. Green, *A Question of Quality: The Higher Education Ratings Game* (Washington, D.C.: American Association for Higher Education-ERIC/ Higher Education Research Report No. 5, 1985).

6. Steven Muller, witness, U.S. Congress, Senate Committee on Appropriations, "Appropriations for HUD and Independent Agencies, 1977." 94th Congress, 2nd Session (1976), p. 1628.

7. Data through June 1985 is included in the study, but due to the way the CIS is compiled, only 12 instances of 1985 testimony were included. The instances of testimony are probably inaccurate, and err on the conservative side. Comparing the CIS Index with other accounts of presidential testimony indicates that the CIS does not always include all witnesses. Nor were all documents always available in the several government depositories accessible to me.

8. Jerome B. Wiesner, witness, U.S. Congress, House Committee on Appropriations "HUD--Independent Agencies Appropriations," 94th Congress, 2nd Session (1976), p. 658.

9. For an articulation of the ideology of pure science see Daniel S. Greenberg, *The Politics of Pure Science* (New York: New American Library, 1967); for accounts of the change to a more applied ideology see David Dickson, *The New Politics of Science* (New York: Pantheon, 1984), and Martin Kenney, *Biotechnology: The University-Industrial Complex* (New Haven: Yale, 1986).

10. David Dickson, *The New Politics of Science*, p. 7. Figures are in 1972 constant dollars.

11. Lawrence Gladieux and Thomas Wolanin, *Congress and the Colleges: the National Politics of Higher Education* (Lexington, MA: Lexington Books, 1976).

12. Dale Corson, witness, U.S. Congress, House Committee on Education and Labor; Subcommittee on Post Secondary Education, "Higher Education Act Amendments of 1976," 94th Congress, 1st Session (July 24, 1975), p. 648.

13. Bruce L. R. Smith and Joseph J. Karlesky. *The State of Academic Science: The Universities in the Nation's Research Effort* (New York: Change Magazine Press, 1977).

14. Dale Corson, witness, U.S. Congress, House Committee on Education and Labor, Subcommittee on Post Secondary Education, "Higher Education Amendments of 1976," 94th Congress, 1st Session (July 24, 1975).

15. As quoted in Corson, "Higher Education Act Amendments of 1976," p. 630.

16. Herbert F. York and G. Allen Greb, "Military Research and Development: A Postwar History," in Thomas Kuehn and Alan L. Porter, eds., *Science, Technology and National Policy* (Ithaca, New York: Cornell University Press, 1981).

17. Jerome B. Wiesner, witness, U.S. Congress, House Committee on Science and Astronautics, "Federal Policy, Plans and Organization for Science and Technology, Part II," 93rd Congress, 2nd Session (June 20, 1974), p. 192.

18. Daniel S. Greenberg, *The Politics of Pure Science.*

19. Donald Hornig, witness, U.S. Congress, Senate Committee on Commerce; Committee on Aeronautical and Space Science, "Science and Technology Applications Act of 1974," 93rd Congress, 2nd Session (July 11, 1974); William D. McElroy, witness, U.S. Congress, House Committee on Science and Astronautics, "Federal Policy, Plans and Organization for Science and Technology, Part II," 93rd Congress, 2nd Session (June 20, 1974).

20. Jerome B. Wiesner, witness, U.S. Congress, House Committee on Science and Technology, Subcommittee on Science, Research and Technology, "Review of the Technology Assessment Act," 95th Congress, 2nd Session (October 20, 1977), p. 454.

21. Richard Lyman, witness U.S. Congress, House Committee on Education and Labor, Special Subcommittee on Education, "Student Financial Assistance/Graduate programs, State Programs and Grants," 93rd Congress, 2nd Session (June 4, 1974), p. 17.

22. Howard Johnson, witness, U.S. Congress, Senate Committee on Labor and Public Welfare, "Postwar Economic Conversion," 91st Congress, 2nd Session (March 23, 1970),p. 438.

23. *Ibid.,* p. 437.

24. *Ibid.,* p. 443.

25. Jerome B. Wiesner, witness, U.S. Congress, House Committee on Appropriations, "HUD—Independent Agencies Appropriations," 94th Congress, 2nd Session (1976), p. 657.

26. A. Bartlett Giametti, witness, U.S. Congress, Senate Committee on Labor and Human Resources, Subcommittee on Education, Arts and Humanities, "Basic Skills, 1979," 96th Congress, 1st Session (February 13, 1979),

from document attached, Giametti, "Sentimentality," *Yale Alumni Magazine*, (January 1976), quotes at pp. 39 and 40.

27. Wesley W. Posvar, witness, U.S. Congress, Senate Committee on Human Resources, Subcommittee on Health and Science Research, "National Science Foundation Authorization Legislation," 95th Congress, 1st Session (March 3, 1977), p. 147.

28. Steven Muller, witness, U.S. Congress, House Committee on Science and Technology, "National Science and Technology Policy Issues, 1979," 96th Congress, 1st Session (April 5, 1979), p. 94.

29. Howard Johnson, witness, U.S. Congress, Senate Committee on Labor and Public Welfare, "Postwar Economic Conversion," 91st Congress, 2nd Session (March 23, 1970), p. 438.

30. David Saxon, witness, U.S. Congress, Senate Committee on Commerce, Science and Transportation, Subcommittee on Science, Technology and Space, "Oversight on OSTP," 96th Congress, 1st Session (March 7, 1979), p. 38.

31. Jerome B. Wiesner, witness, U.S. Congress, Senate Committee on Interior and Insular Affairs, "Energy Research and Development Policy Act," 93rd Congress, 1st Session (June 21, 1973), p. 70.

32. Steven Muller, witness, U.S. Congress, House Committee on Science and Technology, "National Science and Technology Policy Issues, 1979," 96th Congress, 1st Session (April 5, 1979), pp. 94-95.

33. Wesley W. Posvar, witness, U.S. Congress, Committee on Human Resources, Subcommittee on Health and Science Research, "National Science Foundation Authorization Legislation," 95th Congress, 1st Session (March 3, 1977), p. 149.

34. *Ibid.*, p. 147.

35. Wesley W. Posvar, witness, U.S. Congress, Senate Committee on Human Resources, Subcommittee on Health and Science Research, "National Science Foundation Authorization Legislation," 95th Congress, 1st Session (March 3, 1977), p. 149.

36. Stanley Ikenberry, witness, U.S. Congress, Senate Committee on Energy and Commerce, Subcommittee on Health and Environment, "Health Research Act of 1982," 97th Congress, 2nd Session (February 2, 1982), p. 41.

37. Donald Kennedy, witness, U.S. Congress, Senate Committee on Commerce, Science and Transportation, Subcommittee on Science, Technology and Space, "National Science Foundation Authorization," 98th Congress, 1st Session (March 3, 1983), p. 80.

38. John S. Toll, witness, U.S. Congress, Senate Committee on the Ju-

diciary, Subcommittee on Patents, Copyrights and Trademarks, "Uniform Patent Procedures Act of 1983," 98th Congress, 2nd Session (March 27, 1984), p. 87.

39. Harold Brown, witness, U.S. Congress, Joint Economic Committee, "Jobs and Prices in the West Coast Region, 1976," 94th Congress, 1st Session (January 12, 1976), pp. 112-113.

40. For a historical account of the CEOs fascination with the automated "factory of the future," see David F. Noble, *Forces of Production: A Social History of Industrial Automation* (New York: Knopf, 1984). Joe B. Wyatt, witness, U.S. Congress, House Committee on Science and Technology, Subcommittee on Investigations and Oversight, "Computers and Education," 98th Congress, 1st Session (September 28, 1983), p. 67.

41. Richard M. Cyert, witness, US. Congress, Joint Economic Committee, Subcommittee on International Trade, Finance and Security Economics, "Japanese Productivity: Lessons for America," 97th Congress, 1st Session (November 4, 1981), p. 54.

42. David Saxon, witness, U.S. Congress, Senate Committee on Commerce, Science and Transportation, Subcommittee on Science, Technology and Space, "Oversight of OSTP," 96th Congress, 1st Session (March 7, 1979), p. 38.

43. *Ibid.,* p. 35.

44. National Science Board, *Basic Research in the Mission Agencies: Agency Perspectives on the Conduct and Support of Basic Research.* (Washington, D.C.: National Science Foundation, 1978), pp. 57-58.

45. Robert Krinsky, "Swords and Sheepskins: Militarization of Higher Education in the United States and Prospects of its Conversion." *Bulletin of Peace Proposals* 19,No. 1, 1988: p. 37.

46. Committee on the Role of the Manufacturing Technology Program in the Defense Industrial Base, *The Role of the Department of Defense in Supporting Manufacturing Technology Development* (Washington, D.C.: National Academy Press, 1986).

47. David Saxon, witness, U.S. Congress, Senate Committee on Commerce, Science and Transportation, Subcommittee on Science, Technology and Space, "Oversight of OSTP," 96th Congress, 1st Session (March 7, 1979), p. 33.

48. Robert L. Sproull, witness, U.S. Congress, House Committee on Appropriations, "Department of Defense Appropriations for 1982, " 97th Congress, 1st Session (July 29, 1981), pp. 418-419. See also Robert Sproull, witness, U.S. Congress, House Committee on Armed Services, "Hearings on Military Posture," 97th Congress, 1st Session (March 11, 1981).

49. James Olson, witness, U.S. Congress, Committee on Education and Labor, Subcommittee on Post-Secondary Education, "Oversight of Student Financial Aid Programs," 97th Congress, 1st Session (October 15, 1981), pp. 27-28.

50. David Saxon, witness, U.S. Congress, House Committee on Science and Technology, "U.S. Science and Technology Under Budget Stress," 97th Congress, 1st Session (December 10, 1981), p. 149.

51. David S. Saxon, witness, U.S. Congress, House Committee on Armed Services, "Hearings on Military Posture; DOD Authorization for Appropriations for 1983, 97th Congress, 2nd Session (March 2, 1982), p. 1024,1028.

52. Robert L. Sproull, witness, U.S. Congress, House Committee on the Budget, Task Force on Energy and Technology, Task Force on Education and Employment, "Higher Education and Innovation in the U.S. Economy and the President's Fiscal Year 1984 Budget; Perspective from the States," 98th Congress, 1st Session (March 1, 1983), p. 5.

53. *Ibid.,* p. 9.

54. Robert Q. Marston, witness, U.S. Congress, House Committee on Banking, Finance and Urban Affairs; Subcommittee on Economic Stabilization, "Industrial Policy," 98th Congress, 1st Session (June 9, 1983) pp. 24-25.

55. Richard D. DeLauer and Donald Kennedy, *Report of the DOD-University Forum* (Washington, D.C.: Office of the Under secretary of Defense for Research and Engineering, December, 1984), p. 1.

56. *Ibid.,* p. 2.

57. Marvin L. Goldberger, witness, U.S. Congress, Senate Committee on Foreign Relations, Subcommittee on Arms Control, Oceans, International Operations and Environment, "The Future of Arms Control," 97th Congress, 2nd Session (January 20, 1982), p. 91.

58. DeLauer and Kennedy, *Report of the DOD-University Forum*, pp. 1, 3.

59. John Brademas, witness, U.S. Congress, House Committee on Education and Labor, "Oversight on the Impact of the Administration's Fiscal 1986 Budget Proposals on Programs under the Jurisdiction of the Committee on Education and Labor," 99th Congress, 1st Session (February 21, 1985), pp. 366-367.

60. A. Bartlett Giametti, witness, U.S. Congress House Committee on Education and Labor, "Oversight on the Impact of the Administration's Fiscal 1986 Budget Proposals on Programs under the Jurisdiction of the Committee on Education and Labor," 99th Congress, 1st Session (February 21, 1985), pp. 394-398.

61. *Ibid.*

62. John Brademas, witness, U.S. Congress, House Committee on Education and Labor, Subcommittee on Post-Secondary Education, "Hearings on the Reauthorization of the Higher Education Act," 98th Congress, 2nd Session (March 27, 1984), p. 274.

63. Howard W. Johnson, witness, U.S. Congress, Senate Committee on Public Works, Subcommittee on Air and Water Pollution, "National Environmental Laboratories," 92nd Congress, 1st Session (April 28, 1971), p. 494.

64. John S. Toll, witness, U.S. Congress, Senate Committee on the Judiciary, Subcommittee on Patents, Copyrights and Trademarks, "Uniform Patent Procedures Act of 1983," 98th Congress, 2nd Session (March 27, 1984), p. 87.

65. David Dickson, *The New Politics of Science*; Christina Ramirez, "The Balance of Interests Between National Security Controls and First Amendment Interests in Academic Freedom." (Houston: Institute for Higher Education Law and Governance, 1985).

66. Paul E. Gray, witness, U.S. Congress, House Committee on Science and Technology, Subcommittee on Science, Research and Technology, Subcommittee on Investigations and Oversight, "Scientific Communications and National Security," 98th Congress, 2nd Session (March 24, 1984), p. 61.

67. Dale Corson, witness, U.S. Congress, House Committee on Science and Technology, Subcommittee on Science, Research and Technology, Subcommittee on Investigations and Oversight, "Scientific Communications and National Security," 98th Congress, 2nd Session (March 24, 1986), p. 51.

68. DeLauer and Kennedy, *Report of the DOD-University Forum*, Attachment 1.

69. *Ibid.*

70. Richard M. Cyert, witness, U.S. Congress, Senate Committee on Commerce, "Corporate Rights and Responsibilities," 94th Congress, 2nd Session (June 15, 1976), p. 142.

71. Robben W. Fleming, witness, U.S. Congress, House Committee on Education and Labor, Special Subcommittee on Education, "Federal Higher Education Programs, Institutional Eligibility: Civil Rights Obligations," 93rd Congress, 2nd Session (August 12, 1974); James H. Hester, witness, U.S. Congress, House Committee on Education and Labor, Special subcommittee on Education, "Federal Higher Education Programs Institutional Eligibility: Civil Rights Obligations," 93rd Congress, 2nd Session (August 12, 1974); Derek Bok, witness, U.S. Congress, House Committee on Education and Labor, Subcommittee on Post Secondary Education, "Regulations for the Education Amend-

ments of 1976," 94th Congress, 2nd Session (December 16, 1976); Arthur G.Hansen, witness, U.S. Congress, Senate Committee on Labor and Human Resources, "Oversight of the Activities of the Office of Federal Contract Compliance Programs of the Department of Labor," 97th Congress, 1st Session (July 29, 1981).

72. Thomas B. Edsall, *The New Politics of Inequality* (New York: W.W. Norton, 1984); Michael Useem, *The Inner Circle* (New York: Oxford University Press, 1984). For accounts of the history of industry specific regulatory agencies see Gabriel Kolko, *The Triumph of Conservatism: A Reinterpretation of American History* (Chicago: Quadrangle, 1967) and his *Railroads and Regulation, 1877-1916* (New York: W. W. Norton, 1970).

73. Wesley W. Posvar, witness, U.S. Congress, House Committee on Science and Technology, Subcommittee on Science, Research and Technology, "Science and Engineering Education and Manpower," 97th Congress, 2nd Session (February 1982), p. 10.

74. Robert Q. Marston, witness, U.S. Congress, House Committee on Banking, Finance and Urban Affairs, Subcommittee on Economic Stabilization, "Industrial Policy," 98th Congress, 1st Session (June 1983), pp. 23-25.

75. Wesley W. Posvar, witness, U.S. Congress, House Committee on Education and Labor, Subcommittee on Select Education and Postsecondary Education, "Research Needs of Institutions of Higher Education," 98th Congress, 1st Session (December 1983), p. 4.

76. Seymour Martin Lipset, *Political Man* (New York: Doubleday Anchor, 1963).

77. Leo Panitch, "Recent Theorizations of Corporatism," *British Journal of Sociology 31* (June 1980), p. 173.

Chapter 5. Bases of Corporate Interest in Higher Education

1. Business-Higher Education Forum, *An Action Agenda for American Competitiveness* (Washington, D.C.: Business-Higher Education Forum, September, 1986), 37.

2. For a thoughtful discussion of the relationship between pluralism and functionalism, see Patrick Donleavy and Brendan O'Leary, *Theories of the State: The Politics of Liberal Democracy* (New York: Meredith, 1987), pp. 20-22.

3. See, for example, Lynn G. Johnson, *The High Technology Connection: Academic/Industrial Cooperation for Economic Growth* (Washington, D.C.: Association for the Study of Higher Education, 1984, ASHE-ERIC Higher Educa-

tion Research Report No. 6); Thomas W. Langfitt et al., eds. *Partners in the Research Enterprise: University-Corporate Relations in Science and Technology* (Philadelphia: University of Pennsylvania Press, 1983); National Science Board, *University-Industry Research Relationships* (Washington, D.C.: National Science Foundation, 1982); National Science Foundation, *Corporate Science: A National Study of University and Industry Researchers* (Washington, D.C.: National Science Foundation, 1984).

4. See, for example, David Dickson, *The New Politics of Science* (New York: Pantheon, 1984), and Martin Kenney, *Biotechnology: The University-Industrial Complex* (New Haven, CT: Yale University Press, 1986).

5. James Burnham, *The Managerial Revolution* (New York: The John Day Company, 1941); Philip Burch, Jr., *The Managerial Revolution Reassessed: Family Control in America's Large Corporations* (Lexington, MA: Lexington Books, 1972).

6. All told, 76 corporations have participated in the Business-Higher Education Forum since its organization in 1978, but 30 appeared on the roster only once before 1986.

7. *Moody's Bank & Financial Manual* (New York: Moody's Investors Services, 1985), Vol. 1, *Banks, Trust Companies, Savings and Loan Associations, Federal Credit Agencies*, Vol. 2, *Insurance, Finance, Real Estate, Investment Companies*; *Moody's Transportation Manual* (1985); *Moody's Public Utility Manual*, Vols. 1 and 2 (1985).

8. *Council for Financial Aid to Higher Education Casebook* (New York: CFAE, 1984, 13th ed.).

9. Sharon D. Knight, and Deborah Knight, eds., *Concerned Investors Guide: Nonfinancial Corporate Data* (Arlington, VA: Resource Publishing Group, Inc., NYSE, 1983).

10. Linda S. Shaw, Jeffrey W. Knopf, and Kenneth A. Bertsch, *Stocking the Arsenal: A Guide to the Nation's Top Military Contractors* (Washington, D.C.: Investor Responsibility Research Center, 1985).

11. "The Fortune 500," *Fortune* (April 29, 1985): 265-314; "The Service 500," *Fortune* (June 10,1985): 175-202.

12. From this point forward, the discussion of the corporations will rest on *Moody's* since *Fortune* only ranks and does not provide detailed information on the corporations.

13. Neil Fligstein, "The Intraorganizational Power Struggle: Rise of Finance Personnel to Top Leadership in Large Corporations, 1919-1979," *American Sociological Review* 52 (February 1987): 44-58.

14. Interview with Neil Fligstein, author of "Actors and the Ability to Transform Structures: The Case of Diversification of the Largest Firms, 1919–1979," *American Sociological Association (1987), 4282;* and *State and Markets: The Transformation of the Large Corporation* (forthcoming, Harvard University Press).

15. See, for example, Joint Economic Committee, U.S. Congress, *Location of High Technology Firms and Regional Economic Development*, Staff Report prepared for the Subcommittee on Monetary and Fiscal Policy, Joint Economic Committee (Washington, D.C.: USGPO, 1982); National Governors Association, *State Initiatives in Technological Innovation: Preliminary Report of Survey Findings* (Washington, D.C.: National Governors Association, 1983); Office of Technology Assessment, *Technology, Innovation, and Regional Economic Development* (Washington, D.C.: Office of Technology Assessment, 1983).

16. Inadvertently, university presidents participating in the Business–Higher Education Forum may have helped to prevent state encouragement of small, dynamic, high technology corporations. In 1982, Ikenberry of Illinois, Kennedy of Stanford, and Marston of Florida testified against setting aside mission agency research and development funds for small business competitions. The presidents argued that the 3 percent to be taken from the several agencies' budgets was most likely to come out of funds earmarked for basic research, and felt that the basic research budget was already cut too close to the bone. Donald Kennedy and Robert Q. Marston, witnesses, U.S. Congress, House of Representatives, Committee on Science and Technology, "Small Business Innovation Development Act," 97th Congress, 2nd Session, (January 26, 1982); Stanley Ikenberry, witness, U.S. Congress, House of Representatives, Committee on Energy and Commerce, Subcommittee on Health and Environment, "Health Research Act of 1982," 97th Congress, 2nd Session (February 22, 1982).

17. Thus far, I have used Leo Panitch's definition of corporatism as expressed in "Recent Theorizations of Corporatism," *British Journal of Sociology* 31 (June 1980): 159–187, but there are other definitions, less centered on production. See, for example, Philippe C. Schmitter, "Reflections on Where the Theory of Neo-Corporatism has Gone and Where the Praxis of Neo-Corporatism May be Going," in *Patterns of Corporatist Policy-Making*, ed. Gerhard Lehmbruch and Philippe C. Schmitter (London: Sage, 1982): 259-279.

18. I was not able to determine whether the remainder of the corporations were reporting their own research and development efforts separately from their efforts for government. Unless research and development efforts were reported as a separate line on the balance sheet, it was not clear that the two funding sources were being distinguished.

19. National Science Board, *University-Industry Relationships*, 47; see also National Science Foundation, *Trends to 1982 in Industrial Support of Basic Research* (Washington, D.C.: USGPO, 1982), NSF 83-302.

20. Shaw, Knopf, and Bertsch, *Stocking the Arsenal*.

21. Unfortunately, the *CFAE Casebook* is not very clear as to how the corporations included are selected, nor how inclusive its data are.

22. National Science Board, *University-Industry Research Relationships*.

23. The listing of companies and their projects is in Appendix III of National Science Board, *University-Industry Research Relationships*, pp. 137-161. As the introduction notes, the appendix "is not meant to be an exhaustive list of university/industry research interactions, but does contain most of the significant research interactions of these institutions with companies."

24. National Science Board, *University-Industry Research Relationships*.

25. Lists of directors were obtained from the *Moody's* manuals.

26. Philip Caldwell, "The Fiesta Factor,"*Vital Speeches of the Day* 16 (June 1, 1978):510.

27. Board members usually receive remuneration for their services, often between $15,000 and $20,000 per board per annum. When academics sit on several boards, they can come close to earning a second salary.

28. Burnham, *The Managerial Revolution*; see also Burch, Jr., *The Managerial Revolution Reassessed*.

29. Knight and Knight, eds., *Concerned Investors Guide*.

30. The data recorded are for the highest court in which the case has been heard. A number are still in the process of appeal and have not yet been finally disposed.

31. See Thomas B.Edsall, *The New Politics of Inequality* (New York: W.W. Norton, 1984), and Michael Useem, *The Inner Circle* (New York: Oxford University Press, 1984).

32. Articles are cited in the text when a direction quotation is used.

33. Five articles did not fall into the above-mentioned groupings, and were categorized as "other." These included publications in magazines such as *American Legion* and *Futurist*.

34. In the "other" category were 6 articles by five CEO's. Two were very technical accounts of how to build a telecommunicatons network. Two dealt with social themes: one argued that there was hunger in America and spoke for returning to pre-Reagan administration food policies; the other argued for tac-

tical ground zero as essential to meaningful disarmament. Another defended free speech. Yet another was written by a CEO in his capacity as museum president.

35. David M. Roderick, "Tough Enforcement of Trade Laws Urged," *American Metal Market*, 93 (November 12, 1985): 14.

36. Ruben F. Mettler, "Make Trade, Not War!" *Industry Week* 210 (August 10, 1981): 13.

37. Philip Caldwell, "Management Opinion," *Administrative Management* 40 (September 1979): 100.

38. Howard H. Kehrl, "Engineering's Role in National Policy," *Production Engineering* 28 (August 1981): 25.

39. "My Turn," *Newsweek* 93 (April 16, 1979): 16.

40. "The Challenge of Global Competition," *Design News* 41 (July 8, 1985): 26.

41. "Getting Our Share," *Newsweek* 93 (April 16, 1979): 16. Caldwell's words seem to echo those of an auto exeuctive in an earlier era, "Engine" Charlie Wilson (GM), when he said: "What's good for America is good for GM."

42. Edward Donley, "Synfuels, Security and the National Will," *Vital Speeches of the Day* 46 (June 1, 1980): 509.

43. "Seven Wary Views from the Top," *Fortune* 115 (February 2, 1987): 60.

44. David M. Roderick, "Roderick," *Personnel Administrator* 28 (December 1983):72–74.

45. Roger B. Smith, "Outsmarting Competition," *Design News* 42 (February 3, 1986):17.

46. "Seven Wary Views from the Top": 60.

47. Gerald D. Laubach, "Growing Criticisms of Science Impeded Progress," *USA Today* (July 1980):553.

48. "CIA's Inman Focuses Growing Science/National Security Debate," *Chemical and Engineering News* 60 (January 25, 1982):27–29; Adm. Bobby R. Inman, "Classifying Science: A Government Proposal," *Aviation Week and Space Technology* 116 (February 8, 1982):10–11, 82.

49. B.R. Inman, "Guest Commentary," *Design News* 42 (August 18, 1986):17.

50. Donley, "Synfuels, Security and the National Will": 507–509.

51. Edson W. Spencer, "Japan: Stimulus or Scapegoat?" *Foreign Affairs* 62 (Fall 1983):136.

52. Roger B. Smith, "What's Good for General Motors: Liberal Arts," *Across the Board* 23 (May 1986):7–8, 10.

53. Michael Useem, *The Inner Circle* (New York: Oxford, 1984), p. 61.

Chapter 6. The Business–Higher Education Forum Reports: Corporate and Campus Cooperation

1. The reports are *America's Competitive Challenge: The Need for a National Response* (Washington, D.C.: B-HEF, April 1983); *Corporate and Campus Cooperation: An Action Agenda* (Washington, D.C.: B-HEF, May 1984); *The New Manufacturing: America's Race to Automate* (Washington, D.C.: B-HEF, June 1984); *Space: America's New Competitive Frontier* (Washington, D.C.: B-HEF, April 1986); *Export Controls: The Need to Balance National Objectives* (Washington, D.C.: B-HEF, January 1986); *An Action Agenda for American Competitiveness* (Washington, D.C.: B-HEF, September 1986). Two Forum reports are not considered, *The Second Term* (Washington, D.C.: B-HEF, January 1985), and *America's Business Schools: Priorities for Change* (Washington, D.C.: B-HEF, May 1985). The first is not treated because it is a brief assessment of the Reagan administration rather than a policy document. The second is not considered because its focus is specific rather than encompassing higher education as a whole.

2. I have taken this summary of neopluralist positions on the state almost directly from Patrick Dunleavy and Brendan O'Leary, *Theories of the State: The Politics of Liberal Democracy* (New York: Meredith, 1984), especially pp. 288–299.

3. Martin Carnoy, *The State and Political Theory* (Princeton, NJ: Princeton University Press, 1984), p. 250.

4. Ibid., 254. As Carnoy notes, there are other class-perspecitve analyses than the class struggle model, but I have focused on this one because I think it represents the mainstream of neo-Marxian analysis.

5. Philippe C. Schmitter, "Reflections on Where the Theory of Neo-Corporatism Has Gone and Where the Praxis of Neo-Corporatism May be Going," in Gerhard Lehmbruch and Philippe C. Schmitter, *Patterns of Corporatist Policy-Making* (London: Sage, 1982), p. 260.

6. See, for example, work ranging from Thorstein Veblin, *The Higher Learning in America: A Memorandum on the Conduct of Universities by Businessmen* (New York: Viking, 1918), and Upton Sinclair, *The Goosestep: A Study of*

American Education (Pasadena, CA: The Author, 1923), at the beginning of the century, to David N. Smith, *Who Rules the University* (New York: Monthly Review Press, 1974), and Barbra Anne Scott, *Crisis Management in American Higher Education* (New York: Praeger, 1983).

7. For interpretations of business–higher education partnerships centered on high technology, see Robert M. Rosenzweig, with Barbara Turlington, *The Research Universities and Their Patrons* (Berkeley: University of California Press, 1982), especially Chapter 3, "Industry-University Collaboration: The New Partnership"; Lynn G. Johnson, *The High-Technology Connection: Academic/Industrial Cooperation for Economic Growth* (ASHE-ERIC Higher Education Research Report No. 6, Washington, D.C.: Association for the Study of Higher Education, 1984); and Thomas W. Langfitt et al., *Partners in the Research Enterprise: University-Corporate Relations in Science nad Technology* (Philadelphia: University of Pennsylvania Press, 1983). For interpretations which stress training for science-based industry and legitimation, see David F. Noble, *America by Design: Science, Technology, and the Rise of Corporate Capitalism* (New York: Knopf, 1977).

8. The texts are in both pages and columns, and each vary in length. I have indicated when there is deviation from text blocks of 5 inches by 8 1/4 so that the reader can understand how the ratios are calculated.

9. The structure of the argument was determined largely from the synopsis of the arguments, called "executive summaries" or "recommendations," with which all the reports begin.

10. Themes were developed from a content analysis for the entire report. The paragraph was used as the unit of analysis.

11. My analysis of rhetorical structure was informed by Terry Eagleton, *Literary Theory: An Introduction* (Oxford: B. Blackwell, 1983); Fredrick Jameson, *The Political Unconscious: Narrative as a Socially Symbolic Act* (Ithaca, NY: Cornell University Press, 1981), and Gayatri Spivak, *In Other Worlds: Essays in Cultural Politics* (New York: Methuen, 1987).

12. My use of the term *ideology* follows that of Kenneth M. and Patricia Dolbeare in *American Ideologies* (Chicago: Markham, 1971). It is closer to Weber's notion of 'world view' than to Mannheim's 'ideology' versus 'utopia'. More precisely, I mean by *ideology* a fairly coherent set of ideas explaining the problematic issues of the day and offering solutions to them.

13. The disposition of societal resources was determined by sections of the reports that dealt with "recommendations" and "initiatives" which outlined specific actions to be taken in the near future.

14. *America's Competitive Challenge*, iii. Unless otherwise indicated, all quotations in this section are taken from the April 1983 report.

15. For a sense of the conflict and competition that characterized the era, see Hugh David Graham and Ted T. Gurr, eds., *A History of Violence in America* (Englewood Cliffs, NJ: Praeger, 1969), and Robert Wiebe, *The Search for Order: 1877–1920* (Philadelphia: Hill and Wang, 1967).

16. Alfred Lord Tenneyson, *In Memorium*, ed. Susan Shatto and Marion Shaw (Oxford: Clarendon Press, 1982).

17. Labor as an organized entity is not treated fully in the report. Falling levels of unionization are noted with approval, as is unions' acceptance of lower wage rates. However, labor is treated more in terms of the global workforce than in terms of domestic unions.

18. This figure is based on data in the report, which conflicts with figures presented elsewhere. The report says that industry pays for 17 percent of the basic research funded in the United States and that universities perform half of such research, which means approximately 9 percent. However, the figures put forward by the National Science Foundation suggest that the business community provides only about 4 percent of the academic research and development funding, whether for basic research or other.

19. All quotations in this section are taken from *Corporate and Campus Cooperation: An Action Agenda* unless otherwise noted.

20. See note 6, above. Other works that chronicle the tradition of dissent within the academy are Robert Justin Goldstein, *Political Repression in Modern America, 1970 to present* (Cambridge, Schenkman, 1978); Bertell Ollman and Edward Vernoff, eds., *The Left Academy: Marxist Scholarship on American Campuses* (Chicago: McGraw-Hill, 1982); Ellen Schreker, *No Ivory Tower: McCarthyism and the Universities* (New York: Oxford University Press, 1986).

21. The only problem the report envisions in a close relationship between business and industry is conflict of interest over research, especially with regard to publication and free comunication.

22. For an account of how exchange theory works in some academic settings, see Edward T. Silva and Sheila Slaughter, *Serving Power: The Making of the Academic Social Science Expert* (Westport, CT: Greenwood 1984). For dependency theory, see Robert I. Rhodes, ed., *Imperialism and Underdevelopment* (New York: Monthly Review Press, 1970), and James D. Cockcraft, Andre Gunder Frank, and Dale L. Johnson, *Dependency and Underdevelopment* (Garden City, NY: Doubleday-Anchor, 1972). Dependency theory argues that third world nations must be seen in the context of international monopoly capitalism rather than as autonomous nation states. In the context of the university, I see a variant of dependency theory which argues that higher education—even in developed countries—must be seen in the context of international monopoly capitalism rather than as a completely autonomous system.

Chapter 7. The Business–Higher Education Forum Reports: New Issues and Traditional Alliances

1. Business–Higher Education Forum, *The New Manufacturing: America's Race to Automate* (Washington, D.C.: Business–Higher Education Forum, June 1984); *Space: America's New Competitive Frontier* (Washington, D.C.: Business–Higher Education Forum, April 1986); *Export Controls: The Need to Balance National Objectives* (Washington, D.C.: Business–Higher Education Forum, September 1986); and *An Action Agenda for American Competitiveness* (Washington, D.C.: Business–Higher Education Forum, September, 1986).

2. For an analysis of the firms composing the Forum, see Chapter 5, the section entitled "Corporate Investment in Research, Higher Education and University Expertise."

3. All quotations in this section are taken from *The New Manufacturing: America's Race to Automate* unless otherwise indicated.

4. For a historical account of the CEOs' fascination with the automated "factory of the future," see David F. Noble, *Forces of Production: A Social History of Industrial Automation* (New York: Knopf, 1984).

5. Although the factory of the future has no people, it is apparently still gendered.

6. Noble, *Forces of Production*.

7. Frances Moore Lappe, Joseph Collins, David Kimby, and Susan George, *Aid as Obstacle: Twenty Questions About Our Foreign Aid and the Hungry* (San Francisco: Institutes for Food and Development Policy, 1980).

8. Martin Kenney, *Biotechnology: The University-Industrial Complex* (New Haven, CT: Yale University Press, 1986).

9. All quotations in this section are taken from *Space: America's New Competitive Frontier* unless otherwise indicated.

10. Grant McConnell, *The Decline of Agrarian Democracy* (Berkeley: University of California Press, 1953).

11. All quotations in this section are taken from *Export Controls: The Need to Balance National Objectives* unless otherwise noted.

12. Sheila Slaughter, "Academic Freedom and the State: Reflections on the Uses of Knowledge," *Journal of Higher Education* 59 (May/June 1988): 241–262.

13. See, for example, C. Peter McGrath, president of the University of Min-

nesota, witness, U.S. Congress, House of Representatives, Judiciary Subcommittee on Courts, Civil Liberties and the Administration of Justice, "1984: Civil Liberties and the National Security State," 98th Congress, 1st Session (November 2, 1983); Dale R. Corson, president emeritus, Cornell University, and chair for the Government-University-Industry Research Roundtable, and Paul E. Gray, president of MIT, witness, U.S. Congress, House of Representatives, Committee on Science and Technology, Subcommittee on Science, Research and Technology and Subcommittee on Investigations and Oversight. "Scientific Communications and National Security," 98th Congress, 2nd Session (May 24, 1984).

14. Howard W. Johnson, president of MIT, witness, U.S. Congress, Senate Committee on Public Works, Subcommittee on Air and Water Pollution, 92nd Congress, 1st Session (April 28, 1971), p. 494.

15. All quotations in this section are taken from *An Action Agenda for American Competitiveness* unless othewise indicated.

16. Edward T. Silva and Sheila Slaughter, *Serving Power: The Making of the American Academic Expert* (Westport, CT: Greenwood, 1984).

17. Martin Carnoy, *The State and Political Theory* (Princeton, NJ: Princeton University Press, 1984), p. 250.

Chapter 8. Class, State, Ideology, and Higher Education

1. See Chapter 3 for a full discussion of Forum leaders' social-class position.

2. G. William Domnoff, *The Higher Circles* (New York: Random House, 1970).

3. For treatment of the corporations headed by Forum leaders, see Chapter 5. For a discussion of the universities headed by Forum leaders, see Chapter 4.

4. For power elite theories, see C. Wright Mills, *The Power Elite* (New York: Oxford University Press, 1956); G. William Domhoff, *The Powers that Be: Processes of Ruling Class Domination in America* (New York: Vintage, 1978).

5. See James Burnham, *The Managerial Revolution* (New York: The John Day Company, 1941), but see also Philip Burch, Jr., *The Managerial Revolution Reassessed: Family Control in America's Large Corporations* (Lexington, MA: Lexington Books, 1972).

6. See Chapter 3 for a discussion of transinstitutional networks in which Forum leaders participate.

7. Business-Higher Education Forum, *Corporate and Campus Cooperation: An Action Agenda* (Washington, D.C.: B-HEF, May 1984), p. 23.

8. Michael Useem, *The Inner Circle: Large Corporations and the Rise of Business Political Activity in the U.S. and U.K.* (New York: Oxford, 1984), pp.14-15,57.

9. These categories are taken from Thomas R. Dye, *Who's Running America: The Conservative Years* (Englewood Cliffs, NJ: Prentice-Hall, 1986, 4th ed.), p. 12.

10. Ibid.

11. See Ibid. for an explication of the institutionalist position, which sees large institutions and bureaucracy as the source of social and policy power.

12. Business-Higher Education Forum, *America's Competitive Challenge*, p. 27.

13. Ibid., p. 1.

14. For a full discussion of the legislative programs advocated by the Forum, see Chapters 6 and 7.

15. For a discussion of this subject, see Chapters 4 and 7.

16. Barbara and John Ehrenreich, "The Professional-Managerial Class," in *Between Labor and Capital*, ed. Pat Walker (Boston: South End Press, 1979), p. 12.

17. Ehrenreichs, "The Professional Managerial Class," p. 14.

18. Ibid., p. 21.

19. Burton Clark, *The Academic Life: Small Worlds, Different Worlds* (Princeton, NJ: The Carnegie Foundation for the Advancement of Teaching, 1987).

20. Sheldon Krimsky, "The Corporate Capture of Genetic Technologies," *Science for the People* 17 (May/June 1985): 32-37.

21. Martin Carnoy, *The State and Political Theory* (Princeton, NJ: Princeton University Press, 1984).

22. For a critique of the space this question occupies in the debates about the state, see Theda Skocpol, "Bringing the State Back In: Strategies of Analysis in Current Research," In *Bringing the State Back In,* eds. Peter B. Evans, Dietrich Rueschemeyer, and Theda Skocpol (Cambridge: Cambridge University Press, 1985): 3-37.

23. See, for example, Hubert Park Beck, *Men Who Control Out Universities: The Economic and Social Composition of Governing Boards of Thirty Leading American Universities* (Morningside Heights, NY: King's Crown Press, 1947); Morton A. Rauhr, *The Trusteeship of Colleges and Universities* (New York:

McGraw-Hill, 1969); W. H. Cowley, *Presidents, Professors and Trustees* (San Francisco: Jossey-Bass, 1980).

24. Marilyn McCoy and D. Kent Halstead, *Higher Education in the Fifty States: Interstate Comparisons Fiscal Year 1982* (Washington, D.C.: USGPO, 1984, 4th ed.).

25. Robert F. Johnston and Christopher G. Edwards, *Entrepreneurial Science: New Links Between Corporations, Universities and Government* (Westport, CT: Greenwood, 1987); U.S. Congressional Office of Technology Assessment, *Technology, Innovation and Regional Development* (Washington, D.C.: U.S. Congressional Office of Technology Assessment, 1984).

26. Robert Goodman, *The Last Entrepreneurs: America's Regional Wars for Jobs and Dollars* (New York: Simon and Schuster, 1979).

27. For another view of this argument, see Barry Bozeman, *All Organizations are Public: Bridging Public and Private Organizational Theories* (San Francisco: Jossey-Bass, 1987).

28. Business–Higher Education Forum, *America's Competitive Challenge*, p. 3.

29. Business-Higher Education Forum, *Space: America's New Competitive Frontier*, p. 3.

30. For American theorists' views of the state as a site of class struggle, see Carnoy, *The State and Political Theory*, and Martin Carnoy and Henry M. Levin, *Schooling and Work in the Democratic State* (Stanford, CA: Stanford University Press, 1985).

31. Steven Krasner, *Defending the National Interest: Raw Material Investments and U.S. Foreign Policy* (Princeton, NJ: Princeton University Press, 1978).

32. Christopher Jencks and David Riesman, *The Academic Revolution* (New York: Doubleday, 1968). Nonprestigious public institutions were probably not isolated, if academic freedom is any indication. See Walter P. Metzger, "Academic Tenure in America: A Historical Essay," in *Academic Tenure*, Commission on Academic Tenure in Higher Education (San Francisco: Jossey-Bass, 1973).

33. Edward T. Silva and Sheila Slaughter, *Serving Power: The Making of the American Social Science Expert* (Westport, CT: Greenwood, 1984).

34. Jencks and Riesman, *The Academic Revolution*.

35. David Dickson, *The New Politics of Science* (New York: Pantheon, 1984).

36. Lois Peters and Herbert Fusfeld, "Current US University-Industry Research Connections," in National Science Board, *University-Industry Research Relationships* (Washington, D.C.: NSF, 1982).

37. See Chapter 7, where the case of the Indiana partnership is discussed.

38. Peters and Fusfield, "Current US University-Industry Research Connections."

39. William Hartung and Rosy Nimrood, "Pentagon Invades Academia," *Council on Economic Priorities Newsletter* (New York: CEP Publication N86-1), pp. 1-6.

40. Carnoy, *The State and Political Theory*, p. 73.

41. Henry A. Giroux, "Hegemony, Resistance, and Educational Reform," in *Curriculum and Instruction: Alternatives in Education*, eds. Henry A. Giroux, Anthony N. Penna, and William F. Pinar (Berkeley: McCutchan Publishing 1981), p. 418.

42. Business-Higher Education Forum, *America's Competitive Challenge*, pp. 1-2.

43. Business-Higher Education Forum, *Corporate and Campus Cooperation*, p. 26. This statement is an appendix, outlining the "corporate stake" in funding higher education, and therefore represents the position of CEOs rather than of university presidents.

44. Giroux, "Hegemony, Resistance, and Educational Reform," p. 418.

45. For a variety of interpretations of the development of the ideology of expertise, see Mary O. Furner, *Advocacy and Objectivity: A Crisis in the Professionalization of American Social Science, 1865-1905* (Lexington, KY: University of Kentucky Press, 1975); Thomas Haskell, *The Emergence of Professional Social Science: The American Social Science Association and the Nineteenth Century Crisis of Authority* (Urbana, IL: University of Illinois Press, 1977); Dorothy Ross, "The Development of the Social Sciences," in *The Organization of Knowledge in Modern America,* eds. Alexandra Olson and John Voss (Baltimore, MD: Johns Hopkins Univerity Press, 1975): 107-138; Robert Weibe, *The Search for Order, 1977-1920* (New York: Hill and Wang, 1967).

46. See for example Daniel Bell, "The End of Ideology in the West," in *The End of Ideology Debate*, ed. Chaim I. Waxmann (New York: Funk and Wagnalls, 1968): 87-105.

47. Howard Orlans, *Contracting for Knowledge: Values and Limitations of Social Science Research* (San Francisco: Jossey-Bass, 1973); Harland Bloland and Sue M. Bloland, *American Learned Societies in Transition* (New York: McGraw-Hill, 1974).

48. Policy Alternatives for Central America (PACCA), *Changing Course: Blueprint for Peace in Central America and the Caribbean* (Washington, D.C.: Institute for Policy Studies, 1984).

49. Thomas Byrne Edsall, *The New Politics of Inequality* (New York: W.W. Norton, 1984); Peter Steinfels, *The Neo Conservatives: the Men Who Are Changing America's Politics* (New York: Simon and Schuster, 1979).

50. Max Weber, "Class, Status, Party," in *From Max Weber: Essays in Sociology* ed. H.H. Gerth and C. Wright Mills (New York: Oxford University Press, 1946) pp. 193-194.

51. For treatments of various crises of legitimacy in the professions, see note 45.

Index

ABM, 116
Access, to higher education, 30-36, 131-133
   and expansion of higher education, 33-36
   of Blacks, 49, 50, 51
   of Hispanics, 49
   of minorities, 31, 32, 36, 37-40, 50, 51, 113, 131-133
   of poor, 113-133
   of women, 31, 32, 36, 37, 40, 50, 51, 133
   of working class youth, 31, 37-40, 49, 50-51
   to emerging professions, 37-40
   to public sector jobs, 37-40
   to traditional professions, 37-40
America's Competitive Challenge, 52, 174-181, 186, 190, 194, 205
   disposition of societal resources, 180
   ideology, 180-181
   metaphors
      competition, 177-178
      consensus, 177-178
      disordered public policy, 177-179
   structure of argument, 175
   themes
      deregulation, 176
      centralization of government activity, 177
      primacy of business, 177
      privatization, 177
      redistribution, 176
Afghanistan, 200
Agency for International Development, 119
AID, 238
Aid for Dependent Children (AFDC), 40
Allen, Michael Patrick, 81

American Academy of Arts & Sciences, 77
American Association for the Advancement of Science, 29-30, 77
American Association of Community and Junior Colleges, 37
American Association of Higher Education, 10, 77
American Bar Association, 76
American Board of Medical Specialties, 76
American Council on Capital Formation, 243
American Council on Education (ACE), 4, 30, 34, 35, 77, 98, 99, 100, 127, 129
American Economic Association, 76
American Educational Research Association, 77
American Enterprise Institute, 29, 241
American Institute of Chemical Engineers, 76
American Iron & Steel Institute, 76
American Metal Market, 161
American Political Science Association, 76
American Society for Engineering Education, 127
American Statistical Association, 76
America's Watch, 239
Amnesty International, 239
An Action Agenda for American Competitiveness, 205, 206, 204-209
   metaphors
      decline and decay, 206-207
      peril and paralysis, 206-207
   recommendations, 207

INDEX

An Action Agenda *(contd.)*
  structure of argument, 205
  themes
    competitiveness, 205-206
    consensus, 205-206
    governments role in fostering competitiveness, 205-206
    redistribution, 206
Anderson, Robert (Rockwell), 163, 176
Andrew W. Mellon Foundation, 8
Antitrust and regulatory legislation, 52, 135-136, 177
  and Business-Higher Education Forum, 53-54
  and Sherman Act, 136
  cross industry-regulation, 136
  industry specific regulation, 136
Arms Export Control Act, 166
Arthur D. Little, 160
Association of American Rhodes Scholars, 77
Association of American Universities, (AAU), 4, 30, 77, 79, 80, 98, 99, 100, 115, 127, 129
  AAU Presidents testimony before Congress 1970-1985, 97-141
  A Report of the Association of Graduate Schools to the AAU, 115
  Committee on Financing Higher Education, 7
  Committee on Science and Research, 125, 127
Association of Graduate Schools, 127
Astin, Alexander
  Predicting Academic Performance in Colleges, 10
Atomic Energy Commission, 117, 120
Auburn University, 66

Bank America Corporation, 71
Beck, Hubert Park
  Men Who Control Our Universities, 14
Brademas, John (NYU), 131, 132, 133
Brown, Harold (California Institute of Technology), 124
Bohemian Grove Club, 78

Bok, Derek, 14, 43
Bowan, Howard, 10
Bowles, Samuel
  Schooling in Capitalist America, 14
Brookings Institution, 8
Bureau of National Affairs
  Fair Employment Practices, 159
  Labor Relations Manual, 159
Bush, Vannevar, 41
Business-Higher Education Forum, 3, 13, 30, 33, 35, 43, 46, 52, 54, 56, 57, 58, 63, 71, 74, 80, 84, 122, 137, 145, 153, 162, 168, 172, 173, 181, 204, 212, 213
  Careers of members, 69-71
  content analysis of reports, 171-215
  education of members, 65-68
  interlocking directorships shared by members, 71-77
  legislative agenda, 50-55, 137, 138
  members social clubs, 77-78
  membership, 21-22
  military service, 68-69
  origins, 20-21
  proposal for a Bureau of International Competition, 176
  proposal for an Information Center on International Competitiveness, 54, 221
  proposal for a National Commission on Industrial Competitiveness, 54, 176, 221
  position of members in trade and professional associations, 76-77
  position of members on various boards, 72-76
  reports
    An Action Agenda for American Competitiveness, 204, 209
    America's Competitive Challenge, 174-181
    Corporate and Campus Cooperation, 181-187
    Export Controls, 200-204
    The New Manufacturing, 190-194
  similarities and differences with AAU presidents, 136-138
  social location of leaders
Business Roundtable, 4, 29, 73, 82, 83

Index 285

Caldwell, Philip (Ford), 163, 164
California Economic Development Corporation, 75
California Institute of Technology, 67
Carnegie
 Classification of Institutions of Higher Education, 8, 65-68
 Commission on Higher Education, 7, 10
 Corporation, 10
 Council on Policy Studies in Higher Education, 7, 10
 Forum on Education and Economy
 Foundation for the Advancement of Teaching, 72
Carnegie-Mellon University, 83
 Robotics Institute, 124
Carter, Jimmy
 administration, 38, 43-44, 127
Center for the Study of American Business, 241
Central Intelligence Agency, 42, 166
CEOs (Chief Executive Officers), of the Business-Higher Education Forum, 63-87
 address the problem of competitiveness in the global market, 162-163
 address the creation of national consensus, 163-164
 address the factory of the future, 164-165
 address the future of the corporation, 164
 address science and education, 165-168
 and scholarship, 77
 careers, 69-71
 compared to Dye's study of institutional leaders, 80-81
 compared to Useem's study of CEOs, 81-84
 concerns voiced in articles and speeches, 161, 169
 membership in social clubs, 77-78
 membership in trade and professional associations, 76-77
 military service, 68-69
 positions on other corporate boards, 71-72
 position on various boards and social institutions, 72-76
 positions taken in the public press, 160-170
 social composition and origins, 64-65
 speak to problems facing specific industries, 165
 tell success stories, 165
Challenger (space shuttle), 195
Chase Manhattan Bank, 71
Chicago Club, 78
Church Society for College Work, 10
City University of New York (CUNY), 39
Class, *see also* social class, 23, 25
 and a reformulation of neo-Marxist theory, 217-228
 professional-managerial class (PMC), 224-225
 social class location of Forum members, 23, 61-96
Cleveland Foundation
Cohen, Michael D.
 his study of university presidents compared to presidents in the Business-Higher Education Forum, 78-80
Cold War, 41, 42, 117, 119, 134, 204, 221, 238
Cole, Jonathan, 15
Cole, Steven, 15
College Entrance Examination Board, 73
College Retirement Equities Fund, 73
Columbia University, 35
Commission of Independent College and Universities (New York), 39
Committee for Economic Development, 4, 7, 29, 32, 73, 74, 82
Compton, Karl, 41
Computer Assisted Design/Computer Assisted Manufacturing (CAD/CAM), 124
Computer-integrated manufacturing (CIM), 190
Conant, James B., 41
Concerned Investors Guide, 159, 160
Conference Board, 29
Congressional Information Service, 101

Consumer Product Safety Commission, 159
Cornell University, 35
Corporate and Campus Cooperation, 181–187, 174, 205, 212
  ideology: primacy of private sector, 185–186
  metaphors:
    crisis in higher education, 184
    interdependence, 184
  structure of argument, 181–182
  themes:
    antirevisionist interpretation of exchange theory, 182–183, 185
    shared interests, 182–183
Corporation for Science & Technology, 198
corporations, in the Business-Higher Education Forum, 146–158
  dates of incorporation, 146–148
  industrial product lines, 148
  involvement in government research contracts, 152–153
  involvement in high technology, 149
  investment in research, higher education and university expertise, 152, 158
  multinational dimensions, 148–149
  social responsibility, 158–161
  types of corporations, 46
  percentage of sales devoted to R&D, 152–153
  use of academic experts in management, 155, 158
corporatist theory, 4
  and AAU presidents as policy makers, 139–140
  and access to policy making positions, 62–63, 86–87
  and cooperation or shared consciousness between corporate and university leaders, 171, 172, 214–216
  and ideology, 171–172, 214–216
  and the state, 171–172, 214–216
  explanation for bases of cooperation between corporate & university leaders, 143, 145, 149, 170
  interpretation of post-war higher education policy, 58

Corson, Dale (Cornell), 115, 135
CFAE (Council for Financial Advancement of Education), 154
CFAE Casebook, 154
Coordinating Committee for Multilateral Export Controls (CoCom), 200
Council of Graduate Schools, 127
Council on Economic Priorities, 239
Council on Foreign Relations, 73, 74, 82, 83
Council on Post-Secondary Accreditation, 73
cross-industry regulation, 54
  Clean Air Act, 54
  National Environmental Protection Act, 54
  Occupational Health and Safety Act, 54
  Water Pollution Control Act, 54

Danforth Foundation, 10, 72
Department of Agriculture, 113
Department of Commerce, 13, 44, 134, 203
  Business-Higher Education Forum proposal for an Information Center for International Competitiveness in, 55
  Export Administration Regulations, 44
  International Trade in Arms Regulations, 44
Department of Defense (DOD), 30, 41, 42, 43, 44, 74, 75, 113, 115, 117, 120, 121, 125, 127, 128, 129, 131, 133, 137, 138, 145, 153, 203, 204, 231, 232, 233
  as AAU presidents preferred sponsor of research, 120, 121
  Buchsbaum Working Group, 125, 126
  Defense Advanced Projects Research Agency (DARPA), 48, 49, 127
  Department of the Army, 44
  DOD-University Forum, 129, 135
  Office of Naval Research, 49, 120, 124
Department of Defense-Higher Education Forum, 30
Department of Energy, 113, 153, 231, 232
Design News, 161
Detroit Athletic Club, 77

Dickson, David, 16
Domhoff, G. William, 81
Donley, Edward (Air Products & Chemicals), 164, 167
Duquesne Club
Dye, Thomas, 69
  his leadership studies compared to Business-Higher Education Forum CEO's, 80-81, 219

Economic conditions, changing, 28-30
  development of a global economy, 28-29
  falling rate of profit, 29
  growth of high technology, 29
Education
  of Business-Higher Education Forum leaders, 65-69
  and Carnegie Classification, 65-69
Educational Directory of Colleges & Universities, 101
Ehrenreich, Barbara and John, 224
Eisenhower, Dwight D., 116
Environmental Protection Agency, 159, 176
Executive Order Four, 1972, 31
Export Controls, 189, 200, 204
  disposition of societal resources, 204
  rhetorical structure, 202-204
  structure of the argument, 200-201
  themes
    global costs and benefits, 201-202
    independence of non-Soviet bloc nations, 201-202
    nature of science and technology, 201-202
Exxon Education Foundation, 8, 10

Factory of the future, 124, 164-165, 190-191, 192-193
Faculty
  class consciousness, 226
Federal Communications Commission, 136
Federal Reserve Board, 230
Fiscal crisis of the state, 38-40
Flexible Manufacturing Systems (FMS), 124
Food and Drug Administration, 159, 176

Ford Foundation, 10
Ford Motor Company, 71
Foreign Affairs, 161
Fortune 500's, 146, 161, 241
Fox Chapel Club, 78
Fund for the Improvement of Post-secondary Education, 10

General Dynamics, 153
Georgia Institute of Technology, 66
Georgia Science & Technology Commission, 75
GI Bill, 6, 31, 68
Giametti, A. Bartlett (Yale), 119, 132
Gintis, Herbert
  Schooling in Capitalist America, 14
Gladieux, Lawrence
  Congress and the Colleges, 10
Goggin, Malcolm
  Governing Science and Technology in a Democracy, 16
Goldberger, Marvin (Cal Tech), 130
Graduate and Professional Opportunities Program (GPOP), 132-133
Gray, Paul (MIT), 135
Greenberg, Daniel, 16

Hagstrom, Warren, 15
Harvard University, 67, 226, 241
Health and Human Services, 48
Health, Education and Welfare (HEW), 10, 37
Heritage Foundation, 241
high technology, 175, 179
  and the Business-Higher Education Forum reports, 190
  and multinational corporations, 44-45
  and the economy, 28-30
  and university research, 41-45
  military related, 44-45
  resources for, 46-50
Higher education, growth, 32-35, 37-38
  and cheap labor, 34-39
  and state agencies, 37-38
Higher education, retrenchment, 39-40
  and the City College of New York, 39
  and the State University of New York, 39
Hodgkinson, Harold, 10

Hollis, Victor
  Philanthropic Foundations and Higher Education, 14
Hornig, Donald (Brown), 116
Horowitz, Irving Louis
  The Rise and Fall of Project Camelot, 14
Hoover Institute on War, Revolution & Peace, 241

ideology, 23, 25, 235-244
  and the Forum reports, 214-216
  entrepreneurial ideology of expertise, 242-243
  in *America's Competitive Challenge,* 180-181
  in *Corporate and Campus Cooperation,* 185, 186
  in *Space,* 199
  of expertise, 240-242
  social Darwinist, 30-31
Ikenberry, Stanley (Illinois), 122
Illinois Governor's Commission on Science and Technology, 75
Indiana State University, 66
Individual Retirement Accounts, 182
Individual Training Account, 180
Industry and Higher Education, 30
Inman, Admiral "Bobby" (Microelectronics and Computer Technology Corporation), 166-167
Institute of Electrical & Electronics Engineers, 76
Institute for Food and Development Policy, 239
Institute for Policy Studies, 239
  Changing Course Blueprint for Peace in Central America and the Caribbean, 239
institutional class, 51, 52, 219-220, 223, 224, 225, 227
intellectual property, 52, 134
  and Business-Higher Education Forum legislative agenda, 54
  royalties, 123
  patents, 123
Internal Revenue Service, 53

International Council for Equality of Opportunity Principles, 160
Rev. Leon H. Sullivan, 160
Interstate Commerce Commission, 53, 136
Ivy League, 66

Jackson, Jesse
  campaign of, 45
Johns Hopkins University, 35
Johnson, Howard (MIT), 117, 118, 120, 203
Johnson, Lyndon, 116
Justice Department, 159

Karleskey, Joseph
  The State of Academic Science, 15
Kennedy, Donald (Stanford), 14, 122
Kennedy, John F., 116
Kenny, Martin, 16
Krimsky, Sheldon, 16
Kuehn, Thomas
  Science, Technology and National Policy, 16

Langfitt, Thomas W.
  Partners in the Research Enterprise, 13
Laubach, Gerald (Pfizer), 166
Laurel Valley Country Club, 78
Lee, Barbara, 10
Lilly Foundation, 10
Lyman, Richard (Stanford), 117

McElroy, William B. (UCLA)
  Magazine Index, 160
Mansfield Amendment, 42, 43, 121, 125, 126

March, James G.
  study of university presidents compared to presidents in the Business-Higher Education forum, 78-80
Marshall Plan, 238
Marston, Robert Q. (Florida), 129, 137
Massachusetts Institute of Technology (MIT), 117, 118
  Lincoln Laboratories, 233

Index

Massachusetts High Technology Commission, 3
Mellon National Bank, 81, 83
Methodology
 content analysis, 24, 98-111, 172-174
 economic inventories, 24
 for determining the bases of corporate interest in higher education, 145-146
 for locating CEO's remarks in articles, speeches & commentaries, 161-162
 limitations, 25
 power structure research, 23-24, 63-64
 topics & themes in AAU presidents congressional testimony, 112-114
Mettler, Ruben F. (TRW), 163
Militarily Critical Technologies List, 203
Moody's manuals, 146
Morgan Guaranty Bank, 81
Morrill Act, 198
Mortimer, Kenneth, 10
Motor Vehicle Manufacturers Association, 76
Muller, Steven (Johns Hopkins), 119, 120, 121

National Academy of Engineering, 29
National Aeronautics and Space Administration, 117
National Association of Independent Colleges & Universities, 77
National Association of State Universities and Land Grant Colleges, 37, 100, 127, 129
National Bureau of Economic Research, 241
National Council of Churches, 10
National Defense Education Act, 119
National Institute of Education, 10, 37
 Involvement in Learning, 10
National Institutes of Health, 113, 232
National Institute of Independent Colleges, 10
National Labor Relations Board, 145
National Merit Scholarship Corporation, 10
National Research Council, 135
National security
 and ambivalence of AAU presidents to the DOD, 113, 125-131
 and economic prosperity, 113, 130-131
 and increased appropriations for research universities, 113, 130-131
 and research universities during the Vietnam War, 113, 117-121
National Security Council, 230
National Science Board, 144, 152, 155
 University-Industry Research Relationships, 13
National Science Foundation (NSF), 12, 29, 34, 47, 113, 115, 116, 119, 138, 144, 149, 155, 232
National Society of Professional Engineers, 76
National Underwriters, 161
NATO, 200
Nelkin, Dorothy, 16
Neo-marxist theory, 4
 and access to policy-making positions, 62, 86
 and class, 217-228
 and cooperation or shared consciousness between corporate & university leaders, 143-144, 149, 169-170
 and ideology, 171-172, 214-216, 235-244
 and policy making, 97, 139-141
 and the state, 171-172, 214-216
 explanation of bases of cooperation between corporate and university leaders, 143-144, 149, 169-170
 interpretation of post-war higher education policy, 57
Neopluralist theory
 and access to policy making positions, 61, 85
 and cooperation or shared consciousness between corporate & university leaders, 171-172, 214-216
 and ideology, 171-172, 214-216
 and policy making, 97, 139, 141
 and the state, 171-172, 214-216
New Jersey Governor's Commission on Science and Technology, 3

290 INDEX

New Manufacturing The, 189, 190-194
  ideology, 194
  metaphors:
    competition, 191
    displacement of workers, 192-193
    factory of the future, 192, 193
    humans as problem solvers, 192-193
    machines as humans, 192-193
    management & new organizational structures, 192-194
  structure of the argument, 190
  themes:
    centrality of manufacturing, 190-191
    factory of the future, 190-191
    structure of the workforce, 190, 191
New York University, 154
Nixon, Richard, 42, 116
North Carolina State University, 66
1954 Brown Decision, 31
1965 Education Act, 31
Noble, David, 16

Occupational Health & Safety Administration, 159
Office of Education, 10
Office of Strategic Services, 120
Office of Science & Technology, 116
Office of Technology Assessment, 117
Olson, James (Missouri), 127

Pacific Union Club, 77
Peace Corps, 119
Pentagon, 41, 42, 44, 233
P.L. 503, 31
Pledge of Resistance, 239
Pluralist theory, 4
  and access to policy making positions, 61, 85
  and policy making, 138
  elite pluralist theory and policy making, 139
  functional explanation for cooperation between corporate and university leaders, 143, 149, 169-170
  of higher education policy dynamics in the post-war period, 55-56
  of science policy, 56

Policy formation process, national, 1-3
  central actors, 1-3
  new configurations of actors, 1-3
  philanthropic foundations and, 2
  private industry and, 2
  the scientific community and, 2-3
  the state and, 1-2
Policy, higher education
  and war, 41-45
  corporatist interpretation of post-war period, 58
  history of higher education, 27-59
  neo-Marxian interpretation, 57
  pluralist interpretations of post-war period, 55-56
Policy literature in postsecondary education, 5-19
  and the affirmative action literature, 9
  and business, 12-15
  and curriculum, 11-12
  and economic development, 12
  and the independent literature, 9-10
  and the science policy literature, 15-19
  and the state, 11
  and the student aid literature, 8-9
  and the subsidy debates, 7-8
  critical policy literature, 11, 14-15, 16-17
  from a neo-Marxian perspective, 5-12
  from a pluralist perspective, 5
Porter, Alan
  Science, Technology and National Policy, 16
Posvar, Wesley W. (Pittsburgh), 119, 121, 136
President's Commission on Higher Education (Zook Commission), 6
President's Commission on Industrial Competitiveness, 3
President's Science Advisory Committee (PSAC), 116, 117
Privatization, 47-48, 176
Professional-managerial class (PMC), 224-225, 231, 234, 237, 240, 241, 242, 243, 244
Public policy positions, of AAU university presidents, 97-142
Public Utilities Fortnightly, 161

Racquet Club, 77
Reagan, Ronald, 127, 130, 131, 132, 195
 administration, 32, 51, 128, 138, 176, 203, 230, 233
Reams, Bernard
 University-Industry Research Partnerships, 13
redistribution, 176
research, 36
 and the business community, 121-122
 and "grey area", 135
 and high technology, 41-45
 applied, 43
 basic, 43, 44
 classified, 134-135, 203-204
 DNA, 166
 social uses of, 41-45
research contracts
 Celeanese-Yale, 46
 DuPont-Harvard, 46
 Exxon-MIT, 46
 Harvard-Hoescht, 134
 Harvard-Monsanto, 46
 Mallincrodt-Washington University, 46
research pooling agreements
 Microelectronics and Computer Corporation, 53
 Semiconductor Research Corporation, 53
 University Steel Resources Center, 53
Rice University, 66
Ridgeway, James
 The Closed Corporation, 14
Rockefeller Commission, 7
Rockefeller Foundation, 72
Rockwell International, 153
Roderick, David M. (U.S. Steel), 164
ROTC, 42, 128

Salt II, 130
Saxon, David (University of California), 120, 125, 127, 128, 176
Science Advisor to the President, 116
science policy
 corporatist interpretation, 58
 literature, 15-19
 neo-Marxian interpretation, 57

 pluralist interpretation, 56
Scott, Barbara Ann
 Crisis Management in Higher Education, 11, 182
Seven Sisters Colleges, 66
Sinclair, Upton
 The Goosestep, 14
Sloan, Alfred P., 19
Smith, Bruce
 The State of Academic Science, 15
Smith, David N.
 Who Rules the Universities, 14
Smith, Roger B. (GM), 164, 167
social-class, *see also* class
 location of Business-Higher Education Forum members, 61-96
social clubs
 membership of Business-Higher Education Forum members, 77-78
social uses of knowledge, 36-45
 and career preparation, 36-40
 research and service, 41-45
Society of Automotive Engineers, 76
Soviet Bloc, 134, 166
Soviet Union, 200
Space: America's New Competitive Frontier, 189, 195, 199
 disposition of societal resources, 199
 ideology, 199
 metaphors
  competition, 197
  partnership, 197-199
 structure of argument, 195
 themes:
  commercial development of space, 195-197
  potential for research, 195-197
  space and international relations, 196-197
Spencer, Edson W. (Honeywell), 167
Sproull, Robert (Rochester), 127, 128
Stanford University, 241
 Center for integrated Systems, 48
state, 1-2, 23, 25
 and Business-Higher Education Forum legislative agenda, 52, 54-55

292   INDEX

state *(contd.)*
  and corporatist theory, 209-214
  and neo-Marxist theory, 209-214, 228, 235
  and neopluralist theory, 209-214
  argument for the collapse of clear distinctions between the state and civil society, 228-235
  Business-Higher Education Forum and, 209-214
  Business-Higher Education Forum expectations of, 177
Star Wars, 44, 130
State Department, 75
State University of New York, 39
  at Buffalo, 230
status attainment literature
  Wisconsin model, 69
Strategic Defense Initiative (SDI), 16, 45, 130, 135
Sullivan Principles, 161
Supreme Court, 230

tax cuts
  and Business-Higher Education Forum legislative agenda, 52-53
  tax law, 133
testimony of AAU presidents before Congress, themes
  Business higher education partnerships, 113, 121-125
  national security and ambivalence toward DOD, 113, 125-131
  national security and ecomonic prosperity, 113, 125-131
  national security and increased appropriations, 113, 125-131
  national security and the Vietnam War, 113, 117-121
  preservation of access for poor & minority students, 113, 131-133
  preservation of elite status, 113, 114-117
TIAA/CREF, 10
Toll, John (Maryland), 123, 134
Twentieth Century Fund, 72

Union for Radical Political Economics, 239

United Negro College Fund, 73, 154
U.S. House of Representatives, 101
  Committee on Education & Labor, 101-102, 131
    Subcommittee on Postsecondary Education, 111
  Committee on Science & Technology, 111
    Subcommittee on Science, Research and Technology, 111
U.S. Industrial Telephone Association, 76
U.S. Office of Technology Assessment, 13
U.S. Senate, 101
  Committee on Commerce, Science & Transportation, 111
  Committee on Foreign Relations, 111
  Committee on Labor and Human Resources, 111
  Committee on Labor & Public Welfare, 111
    Subcommittee on Science, Technology & Space, 125
U.S. Steel Foundation, 8
University-industry agreements
  Case Western Reserve Center for Applied Polymer Research, 47
  University of Delaware Center for Catalytic Technology, 47
  University of Massachusetts Center for Research on Polymers, 47
University-Industry-Government Roundtable, 30
University of Buffalo, 230
University presidents, 63-87, 97-141
  AAU presidents
    and public policy, 97-141
    testimony before congress, themes. *See also* testimony before Congress, themes, 114-133
  Business-Higher Education Forum presidents
    and scholarship
    education of, 65-68
    careers, 69-71
    compared to Cohen and March's study of president's, 78-80
    education of, 65-68
    membership in social clubs, 77-78

membership in trade and professional associations, 76–77
military service, 68–69
positions on corporate boards, 71–72
positions on various boards and social institutions, 72–76
social composition, 64–65
Useem, Michael
    his study of CEO's compared to Business-Higher Education Forum CEO's, 81–84, 219

Veblen, Thorstein
    The Higher Learning, 14, 182
Vietnam War, 41, 42, 45, 113, 114, 116, 117–121, 125, 126, 131, 140, 232
Vital Speeches, 161

Washington University, 241
Watergate, 121

Weber, Max, 241
Westinghouse, 124
Wiesner, Jerome (MIT), 113, 116, 117, 118
Wilson, John T.
    Academic Science Higher Education and the Federal Government, 18
Witness for Peace, 239
Wolanin, Thomas
    Congress and the Colleges, 10
Wolfle, Dael
    Science and Public Policy, 15
Woodside, William S. (American Can), 164, 165
World War II, 31, 36, 41, 46, 52, 116, 118, 120, 121, 124, 130, 189, 201, 213, 231
Wyatt, Joe B. (Vanderbilt), 124

Zook Commission (President's Commission on Higher Education), 6, 7